Exploring Tidepools

This is a book about exploring the tidepools that dot the southern California coast from San Luis Obispo County to the Mexican border (see map, page 47). It is about the kinds of things you may find in these pools, and about the best way to observe these plants and animals without hurting them.

People used to think that there is so much life in tidepools that you could never collect so many animals that they would become endangered. They used to think that the ocean was so big that you could never pollute that much water. Now we know that both of these ideas are wrong. Today many tidepool plants and animals are rare, and one of the main reasons for their scarcity is pollution. So, when you visit the tidepools, we urge you to be careful and considerate. Don't pull animals off the rocks, or poke them with sticks, or deliberately step on them. Remember that they are living things that share our planet with us, and that people who come after you will want to see them too.

Using this guide, we hope you will learn to watch tidepool life right where it is, learn from it, enjoy it, and leave it there.

We hope you'll have a lot of fun, too!

Contents

Activities

SANTA BARBARA MUSEUM OF NATURAL HISTORY

Tidepool Coloring Page

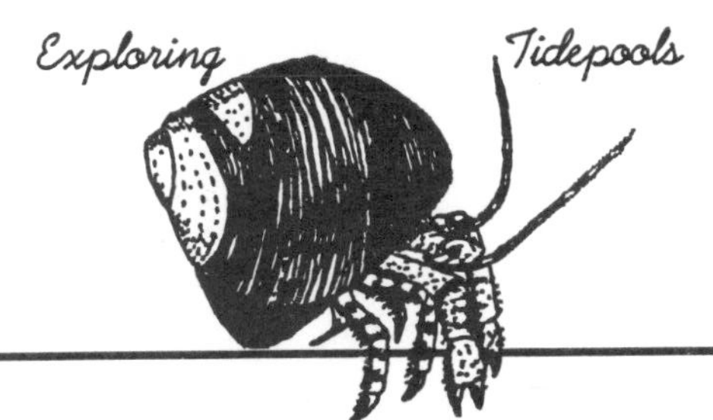

What Are Tidepools?

True to their name, tidepools are pools left behind when the water recedes at low tides. These pools may be trapped in the hollows of rocks, between rock piles, or sometimes in sand. Lots of plants and animals can be seen in these pools. Some of them live there all the time but you can only see them during low tides, while others are trapped there by the drop in water level. It is these plants and animals that make exploring the tidepools an adventure. Since tidepools are covered over by water during each high tide, new water, new plants and animals can get in while others may get out. So you could visit the same tidepool many times and see something different each time!

Be a Tidepool Explorer!

Dress the part.

A good tidepool explorer is dressed for the experience:

A **HAT** is important for protection from the sun. It also helps you to see better into the pools.

A **LIGHT COTTON SHIRT** or T-shirt will dry quickly if it gets wet.

SHORTS or a swim suit.

STURDY SHOES that are OK to get wet.

A light, **WINDBREAKER-TYPE JACKET** is a good thing to bring along too, since the wind is sometimes cool or you may get cold after being wet.

It's also a good idea to bring along a towel and extra shoes and socks.

Your **TIDEPOOL EXPLORER'S KIT** (see next page).

Tidepool Etiquette

Etiquette (ET-te-ket) means *proper manners and behavior.* If you go to visit someone at their home, you are expected to behave properly. When you visit tidepools, you are visiting the home of many little plants and animals, and you should behave properly there, too.

Be Careful. This applies not only to you, so you don't hurt yourself, but also to the animals. You will have to be careful not to step on creatures, and to walk carefully over the rocks and sand.

Don't Pry. A lot of animals, such as Sea Stars and Anemones, attach themselves firmly to rocks. Don't try to pull or pry them off -- you will tear parts of their bodies and hurt them.

Turn Rocks Carefully. Lots of things live under rocks. Some rocks are small enough for you to turn. Do so slowly and carefully, look at what you find, and then *carefully return the rock to its original position.*

Don't take anything living away. Seashore life can't live away from the seashore! It is OK in *some* (not all) areas to collect empty shells, but never take living things away from the tidepools.

Don't poke or prod. Many tidepool animals are soft and delicate. Some may be touched gently with a finger, but others should not be touched at all. *None* of them should be poked with sticks or other hard objects.

Be aware of the laws. Most tidepool life in California is protected by law. Many tidepools are in parks and preserves, where even empty shells are protected. Make sure you know the laws for the area you visit.

The Tidepool Explorer's Kit
Some stuff you might bring along to help you explore tidepools!
HIGH LOW TIDES 1993
Tide Book
(see page 9)
Aquarium Dip Net
You can get this at a tropical fish store. Be sure to get the kind with plastic wrapped around the handle -- wire will rust fast in salt water!
Use it to gently scoop out little fish, shrimps, etc. to get a better look!
Magnifying Glass
You'll see pictures throughout this book on how this will be handy!
Day Pack
Use this to carry all your other things!
Plastic Pail
Use this to look more closely at small animals you catch and to carry them back to their homes after you do!
15
Plastic Jar
Lets you see tiny animals more easily
Sunscreen
SPF 15 minimum is recommended
NEWS-PRESS
Waterscope
(See Page 12)
Newspapers
To wrap up empty shells to bring them home
(see page 37)
Plastic ("zipper") Bags
Good for carrying things and seeing them better. But don't leave them at the beach (see page 45)

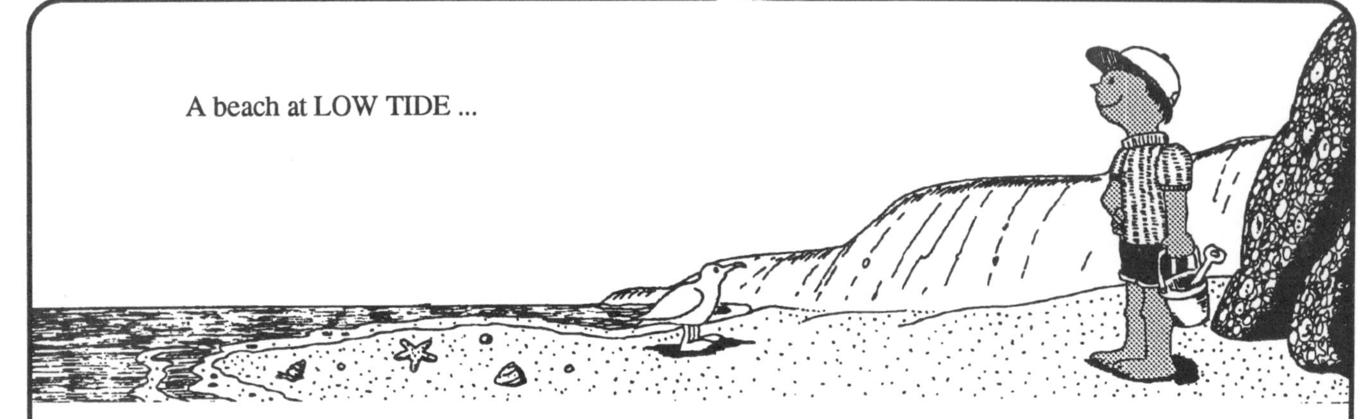
A beach at LOW TIDE ...

What Are Tides?

Every day the water of the ocean washes up higher and higher onto the shore until it reaches a certain point; then it begins to recede, until it reaches a low point. These high and low points occur about six hours apart, so unless you were on the beach for a long time you might not notice that the water level changes much. You'd see that some waves are bigger than others, but waves have nothing to do with the tides. The tides are the steady rising and falling of the water level each day.

... the same beach SIX HOURS LATER!

Tides are caused by the *gravitational pull* of the earth, moon and sun upon each other. The moon has a greater effect on the tides because it is closer to the earth. Since water is more "flexible" than land, it actually bulges up in the direction of the moon, as shown in the diagram below (but it wouldn't look like this from outer space; the layer of water on the earth is not thick enough to *really* see a "bulge").

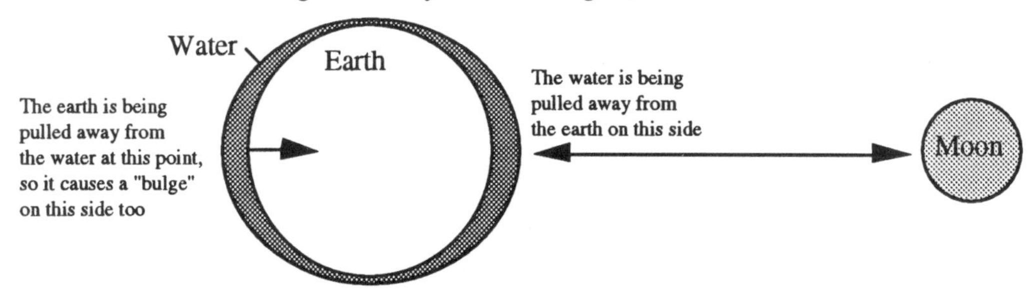

Spring Tides and Neap Tides

Sometimes the sun and the moon will be in a line with the earth, and during these times (which are also the times of the *full* and *new* moon) the tides reach their highest highs and lowest lows of each month. These are called *spring* tides. This has nothing to do with the season we call spring, since they happen about twice each month. It means "to leap or jump."

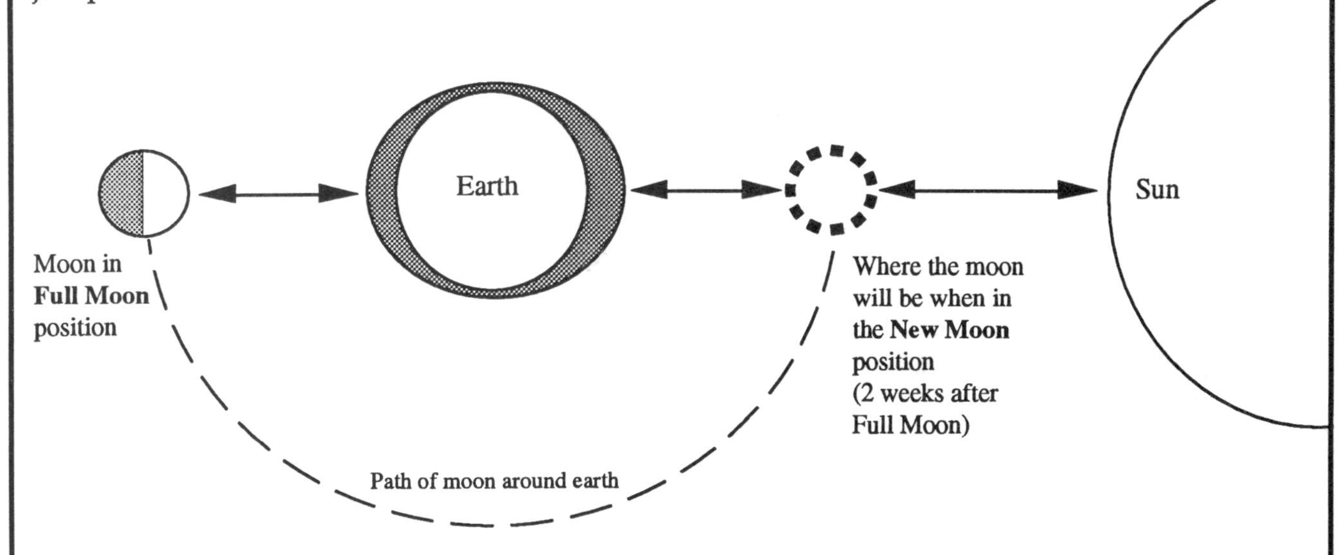

During the times of the first quarter and last quarter moon, the moon and sun are at right angles to the earth, and their gravitational pulls sort of balance each other. High tides aren't as high, and low tides not as low as during spring tides. These are called *neap* tides.

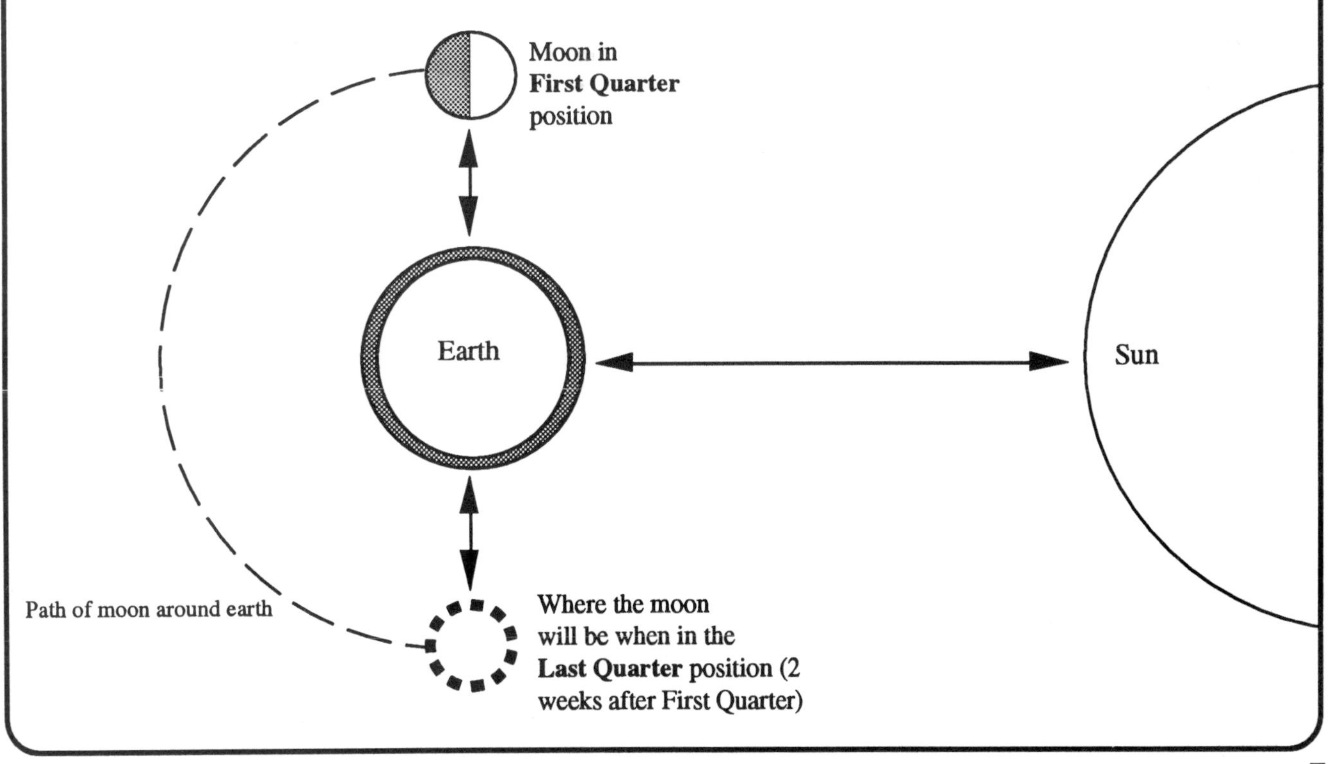

Whirling Along Beneath the Tides

The water isn't *really* going up and down. It is always bulging toward the moon (and sun), but the earth is rotating, making a complete turn in 24 hours.

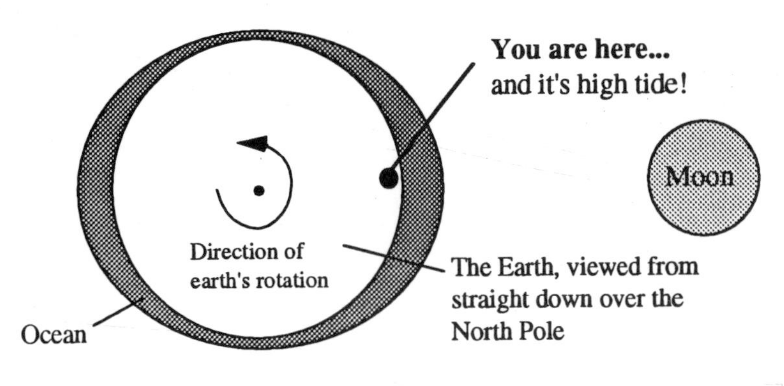

So, if you are at a certain spot on the earth and it happens to be right under the "bulge," then it is high tide where you are.

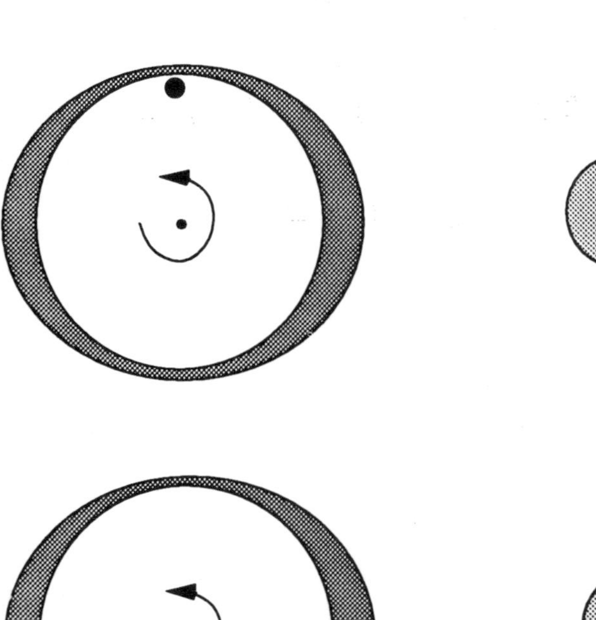

Six hours later, the earth will have rotated 1/4 turn, and you will be on the side of the earth where the water is lower and it is low tide.

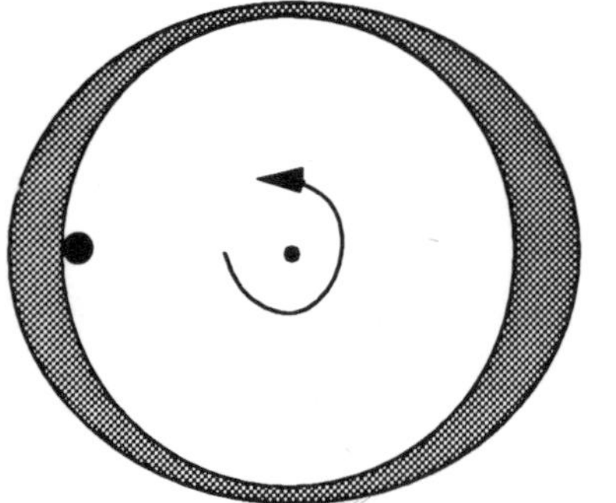

In another six hours, you will be under the bulge on the opposite side of the earth -- high tide again!

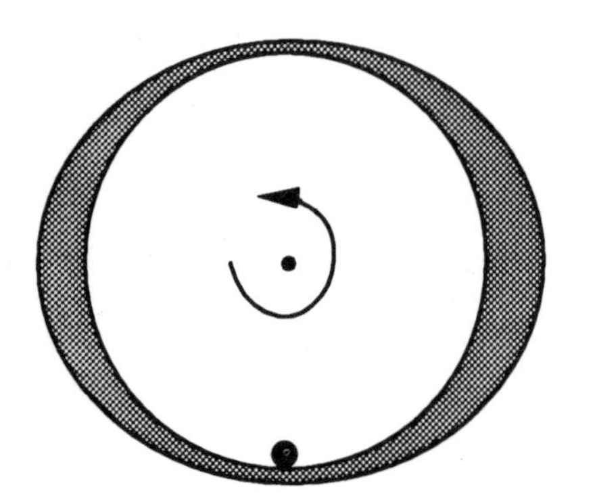

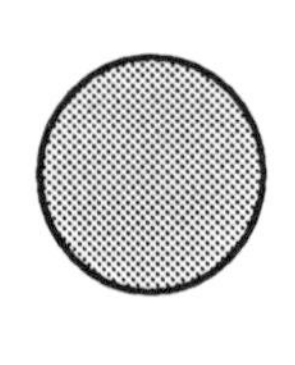

In yet another six hours, you'd be under shallow water, at your second low tide for the day.

Finally, at the end of the 24 hours, you are back where you started in high tide. But during that time, the moon has moved about 1/28th of the way in its orbit of the earth, so it really takes about 24 hours and 50 minutes to reach the "starting point" again. That's why the high and low tides occur at different times each day.

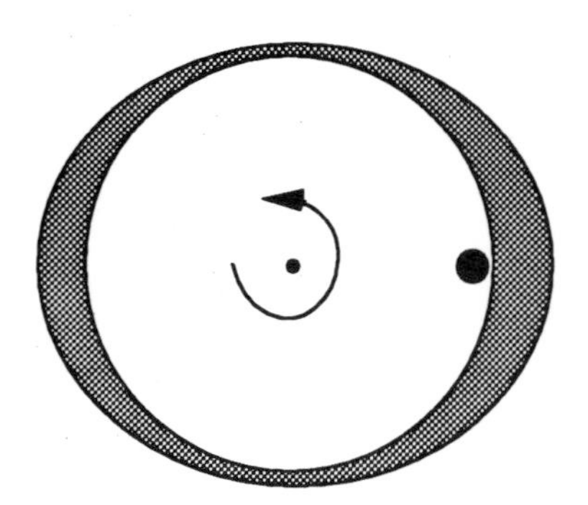

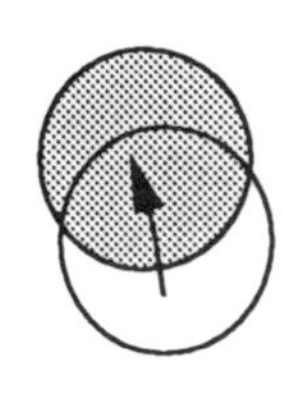

Finding the Tide Schedules in Your Area

On the west coast, we have a complicated *mixed semi-daily* tide pattern. That means that we have about two high and two low tides each day, but they are not equal. One high is higher than the other, and one low is lower.

You can find the tides for the day or the week listed in some newspapers, or you can get a small book with listings for the year from fishing tackle stores, bait shops, boat landings, natural history museums and marine aquariums. When planning a visit to the tidepools, look for low tides that are listed as being below zero feet. These are called *minus* tides, and have a minus sign (-) in front of the number that tells how many feet below zero they are. Minus tides are printed in red in those little books mentioned above. The higher the number of feet below zero a tide is, the farther out the water will be and the more marine life will be exposed. These are the best times to visit tidepools. In southern and central California, the best minus tides occur during the day in the winter months.

Intertidal Zones

Some parts of the beach are higher than others, and some rocks are higher than others. This means that the water reaches some parts during high tides, but not others; and some parts may be below the water except during minus tides. This creates four *zones,* or regions of the beaches or rocks that are affected by the water in different ways. We call these *intertidal zones.* They are illustrated below.

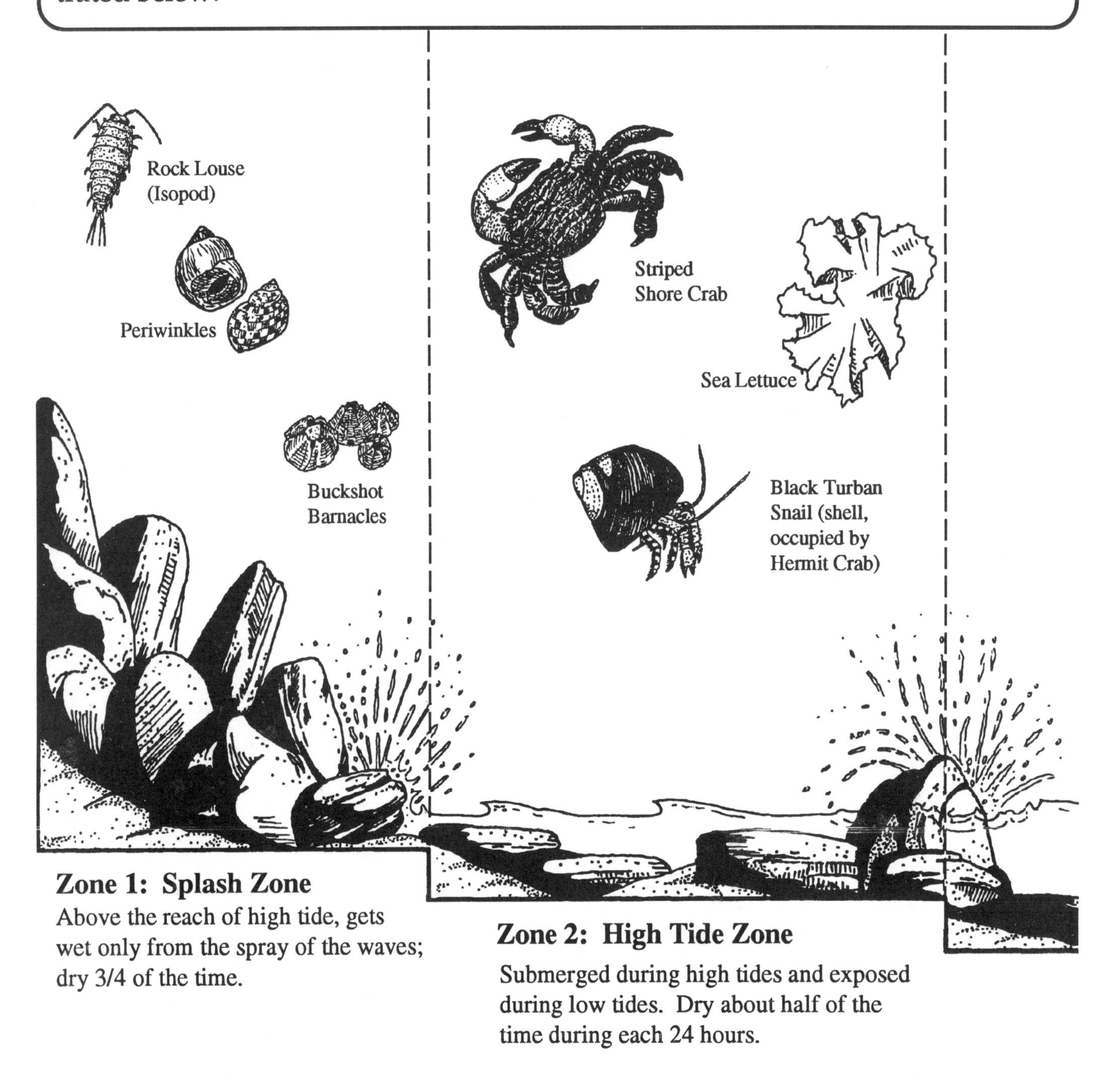

Zone 1: Splash Zone
Above the reach of high tide, gets wet only from the spray of the waves; dry 3/4 of the time.

Zone 2: High Tide Zone
Submerged during high tides and exposed during low tides. Dry about half of the time during each 24 hours.

Animals and plants of the seashore are adapted to life in one or more of these zones. Which zones an animal can live in depends on how well it can survive certain conditions. For instance, an animal in the **High Tide Zone** must be able to live through being dry half the day. This also means being exposed to the heat and light of the sun, drying out, the pounding of waves, increased salt (because only water evaporates — the salt gets left behind and *concentrates* in the tidepools), and being exposed to predators such as birds and other land animals.

Some animals or plants are found in only one zone. When you find that animal, you know you are in that zone. These are called *indicator species* because of this. For instance, the **Rock Louse** is an indicator of the **Splash Zone.**

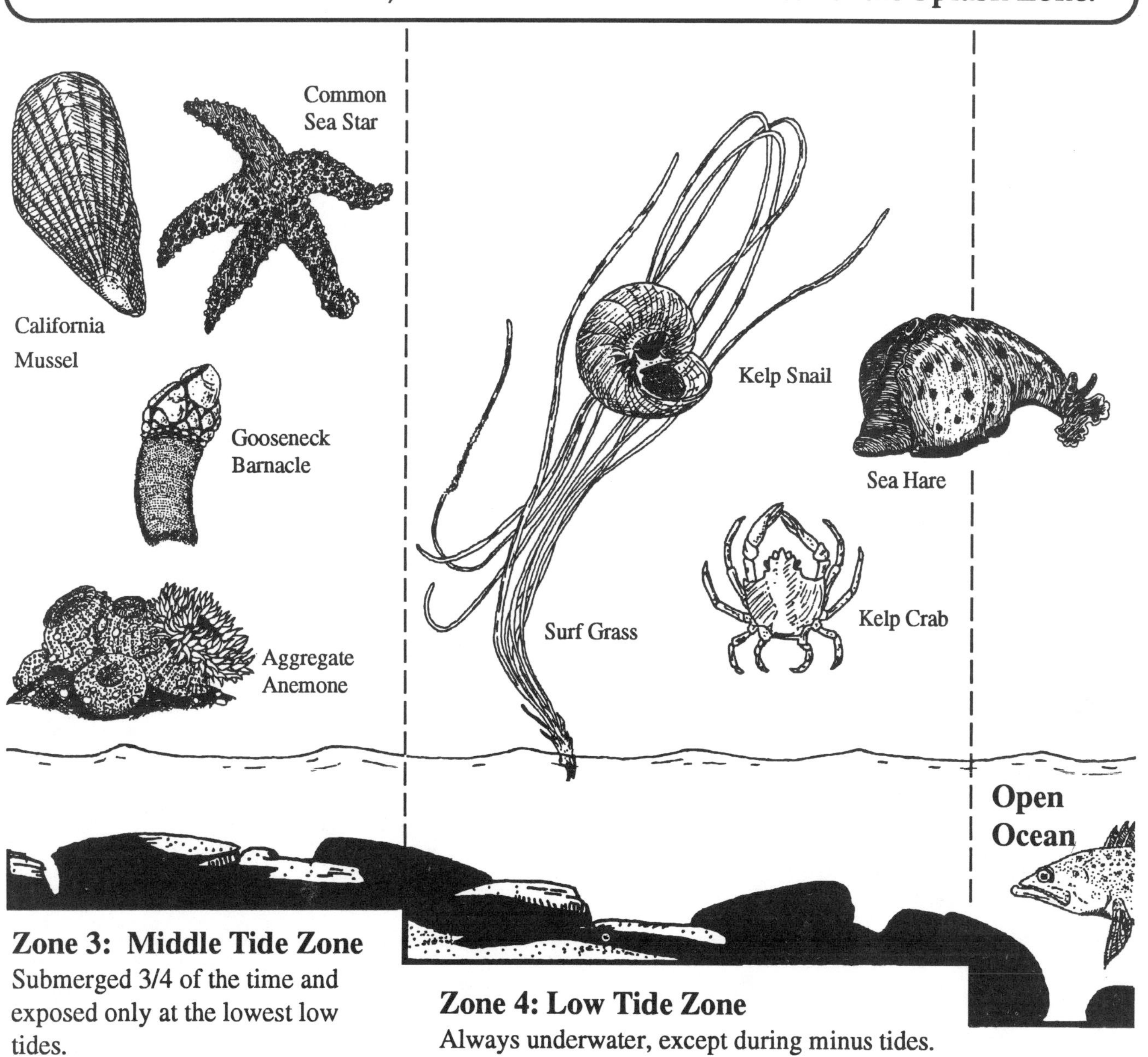

Make a WATERSCOPE!

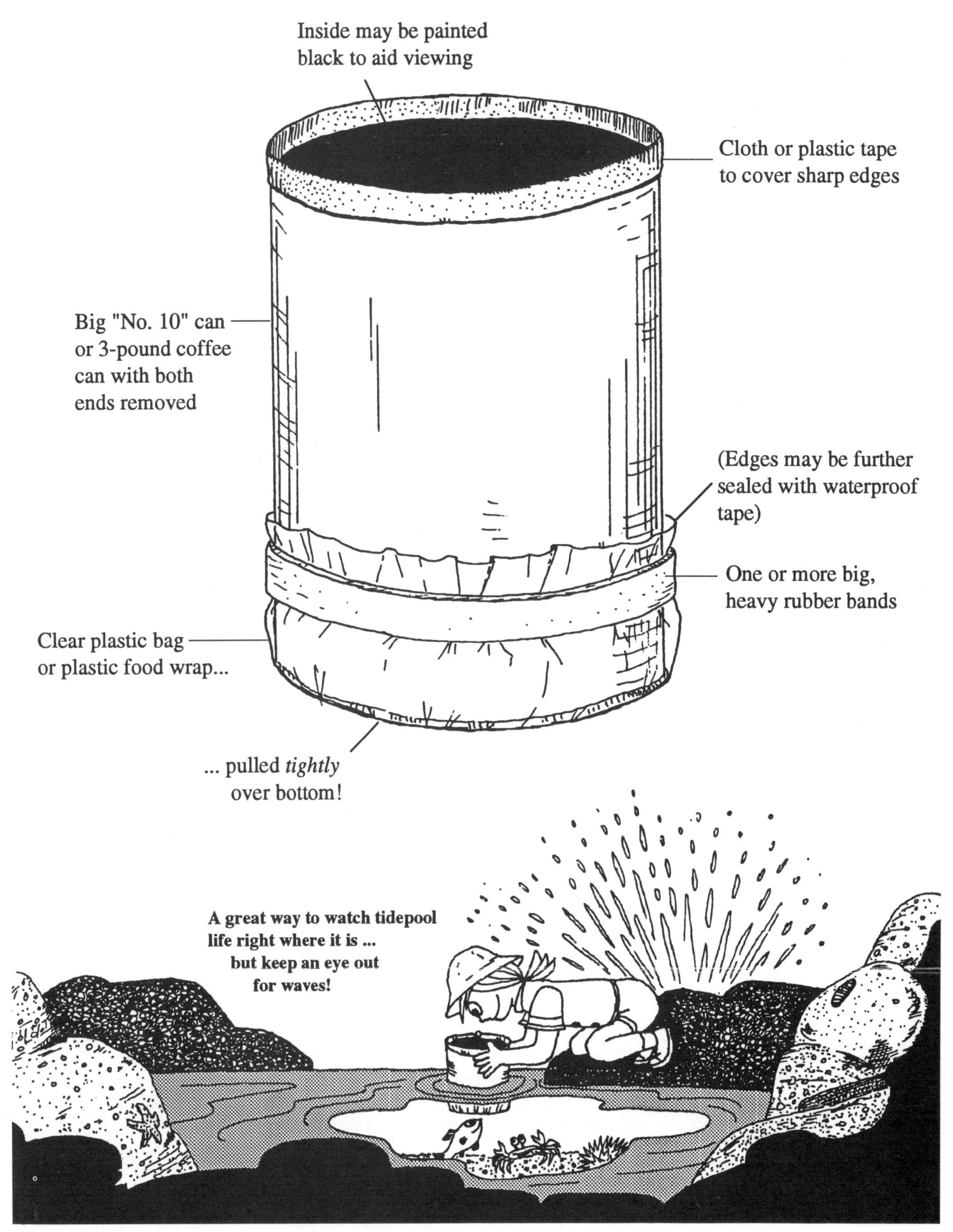

Classification

In order to better understand and study living things, scientists group them together, or *classify* them. They group them together by the *characteristics* that they share with one another. For instance, crabs, lobsters, shrimp and barnacles all have jointed legs, so they are all classified together. The major groups, called *phyla* (singular: phylum) of living things that you might see in tidepools are shown below:

Plant Kingdom:

Algae (AL-jee). Most marine plants belong to one of three phyla of algae:

Flowering Plants. These plants have true leaves, roots and stems. These are the most common plants on land, but they are rare in the sea.

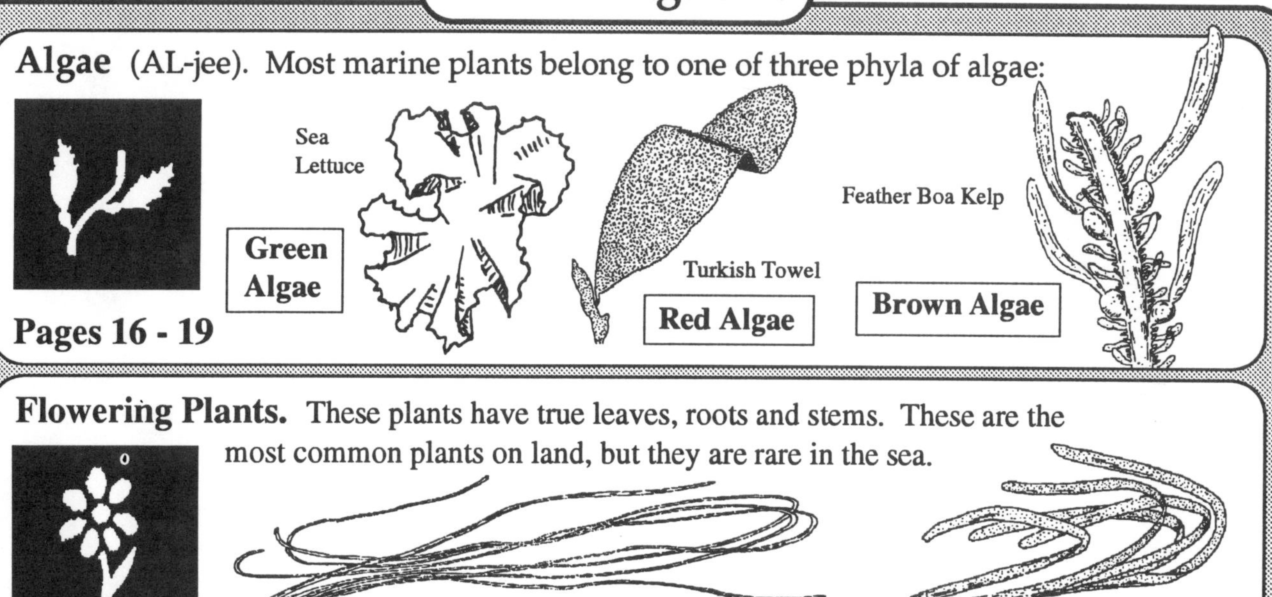

Pages 16 - 19

Page 20

Surf Grass and **Eel Grass** are two common marine flowering plants.

Animal Kingdom:

Phylum

Porifera (pore-IF-fir-ah). Sponges. Usually a bumpy or sheet-like growth on rocks or other objects, with a spongy or gritty feel to it. Sometimes many small holes can be seen.

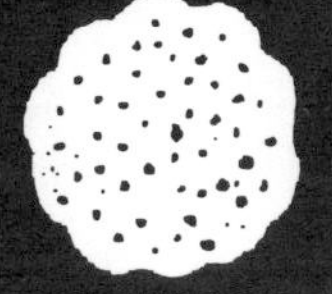

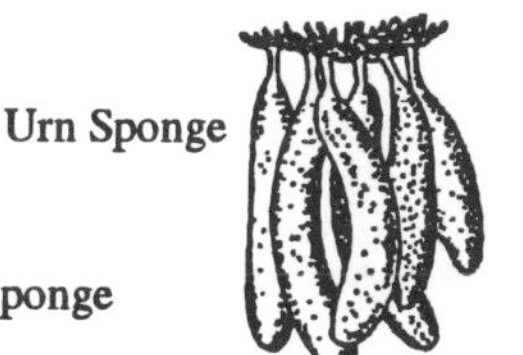

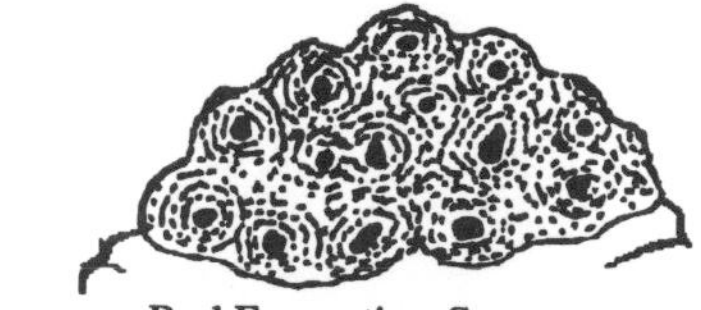

Page 23

Phylum

Cnidaria (nye-DARE-ee-ah). "Stinging animals." Hollow-bodied animals with stinging cells for catching food and for protection.

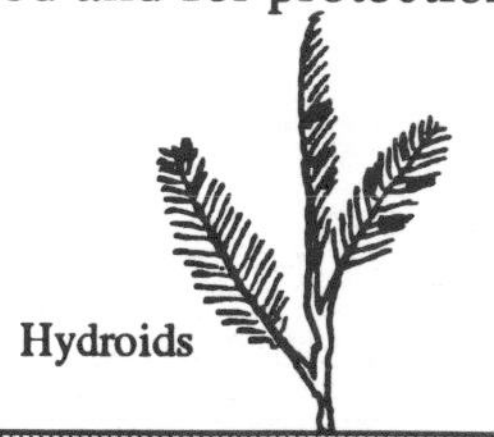

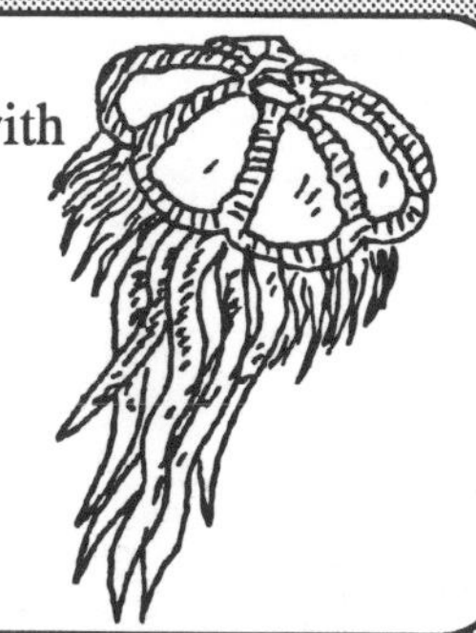

Pages 24 - 25

Animal Kingdom, continued

Phylum

Bryozoa (brye-oh-ZOH-ah). "Moss animals." Very tiny animals found in colonies. Although often common, they are usually overlooked.

California Moss Animal

Jack Frost Bryozoan

Page 20

Marine Worms: There are actually several phyla of worms that occur in tidepools.

Peanut Worms

Ribbon Worms

Flatworms

Segmented Worms

Pages 26 - 27

Phylum

Arthropoda (ar-THROP-oh-dah). "Joint-legged animals." Animals with *exoskeletons* and jointed legs.

Isopod

Hermit Crabs

Barnacles

Shrimps

True Crabs

Pages 28 - 30

Phylum

Mollusca (maw-LUS-kah). "Soft-bodied animals." Many of these animals protect their soft bodies with hard shells.

Chitons

Limpets and Snails

Sea Slugs

Octopus & Squid

Clams, Mussels, etc.

Pages 32 - 36

Phylum

Echinodermata (ee-KINE-oh-der-MAH-tah). "Spiny-skinned animals." Animals with bodies arranged in five parts, and with "tube feet."

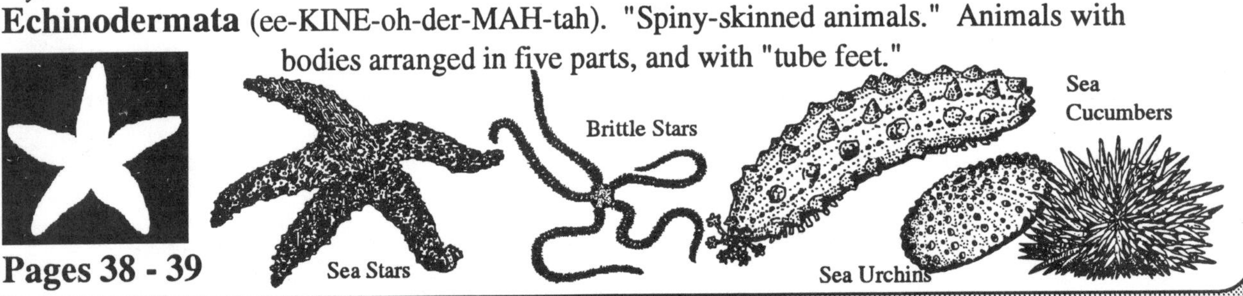

Pages 38 - 39

Phylum **Chordata** (kor-DAH-tah). Animals with a *notochord* and/or backbone.

Subphylum

Tunicata (too-nih-CAH-tah). Tunicates or "sea squirts." These unusual animals are distantly related to backboned animals.

Page 41

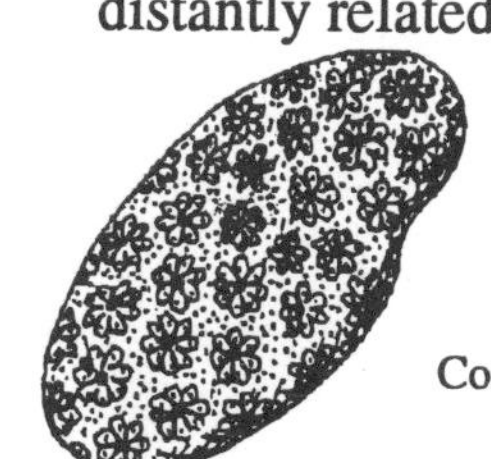

Compound Ascidian

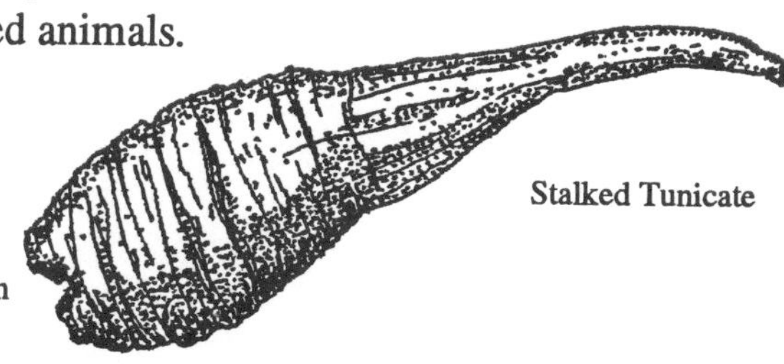

Stalked Tunicate

Subphylum

Vertebrata (ver-teh-BRAH-tah). Animals with backbones. There are two major vertebrate groups seen at tidepools - birds and fishes. In this book we only consider fishes.

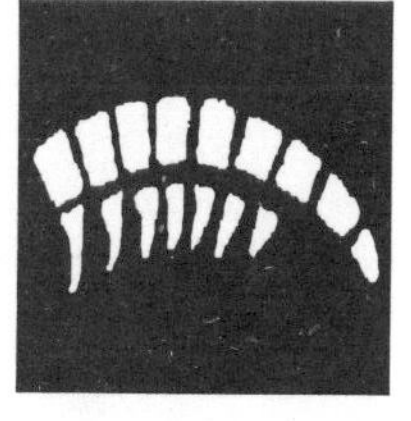

Page 42

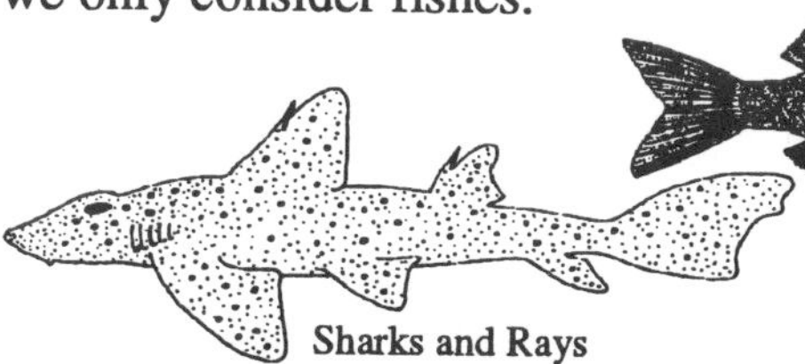

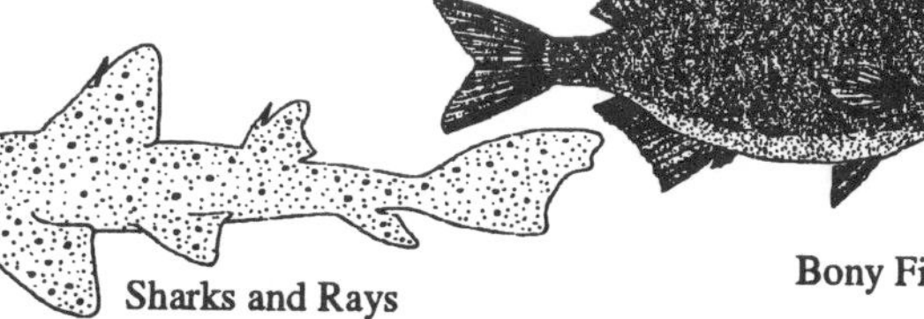

Sharks and Rays

Bony Fishes

Why two names?

Every plant and animal illustrated in this book has two names listed next to it. The first is its *common name,* or the name that most people call it. Nearby is the *scientific name,* the name it is known by to scientists. Common names are all right for everyday use, but they don't work in all situations. For instance, what we call a "Common Sea Star" refers to the common sea star of the Pacific Coast of the United States. But it isn't the same as what someone on the Atlantic Coast would call the common sea star, or someone in Peru or someone in Africa. So *common names* can be confusing!

But *Pisaster ochraceus* is the name of just one kind of sea star in the world, and that's the one that is found along our coast. So if a scientist in Peru or Africa or Massachusetts is talking about *Pisaster ochraceus,* he or she is talking about "our" Common Sea Star (which is also sometimes called the Ochre Sea Star, another example of how common names can be confusing)!

(Some people call sea stars "starfish," but they really aren't fish at all, so we prefer to say sea stars.)

Seaweeds

California waters are home to more than 600 hundred kinds of marine *algae* (AL-jee), commonly called "seaweeds." They are classified into three phyla, based on their color: Red, brown or green.

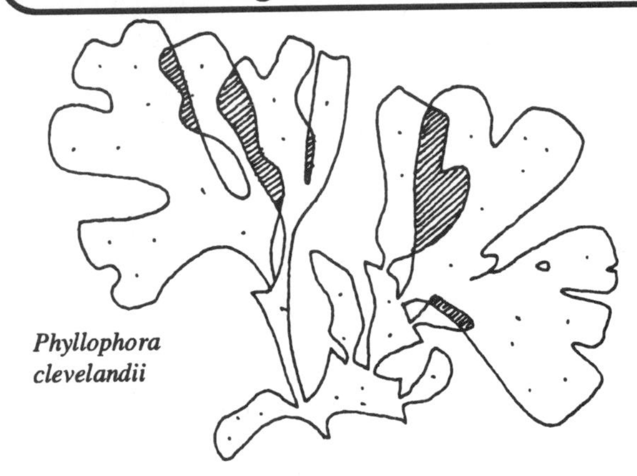

Phyllophora clevelandii

Phyllophora (FILL-oh-FOR-ah)
Brownish-red, a deep-water species that sometimes washes up into tidepools

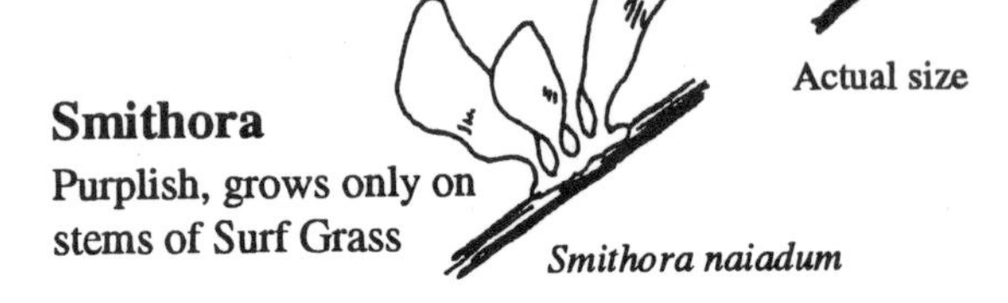

Smithora
Purplish, grows only on stems of Surf Grass

Smithora naiadum

Red Algae: Phylum **Rhodophyta**
(ROAD-oh-fie-tah)
Red algae are the most abundant seaweeds. There are more than 4000 kinds worldwide. They are various shades of red, pink, purple, reddish-brown or even greenish, and many are difficult even for experts to identify.

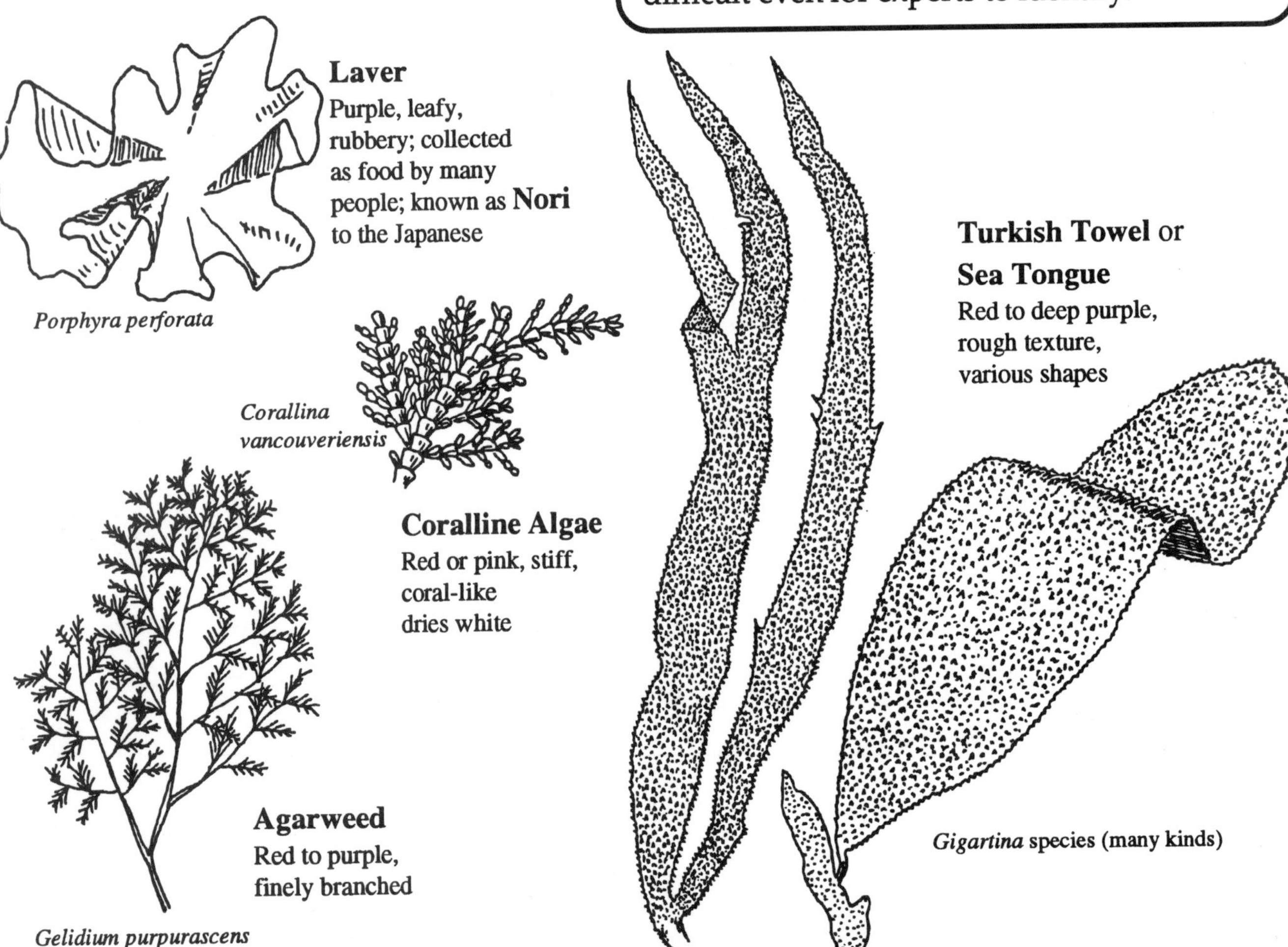

Laver
Purple, leafy, rubbery; collected as food by many people; known as **Nori** to the Japanese

Porphyra perforata

Corallina vancouveriensis

Coralline Algae
Red or pink, stiff, coral-like dries white

Turkish Towel or **Sea Tongue**
Red to deep purple, rough texture, various shapes

Gigartina species (many kinds)

Agarweed
Red to purple, finely branched

Gelidium purpurascens

Algae are not flowering plants, and their structures are different from the parts of flowering plants. So, they have different names. Instead of roots, some marine algae have a *holdfast* which looks like a mass of roots, but as its name implies is used to "hold fast" to some object, such as a rock on the bottom. What looks like a stem is called the *stipe*. And what you might call a leaf is properly called a *blade*. Some seaweeds have little gas-filled bulbs, called *bladders*, that help them float.

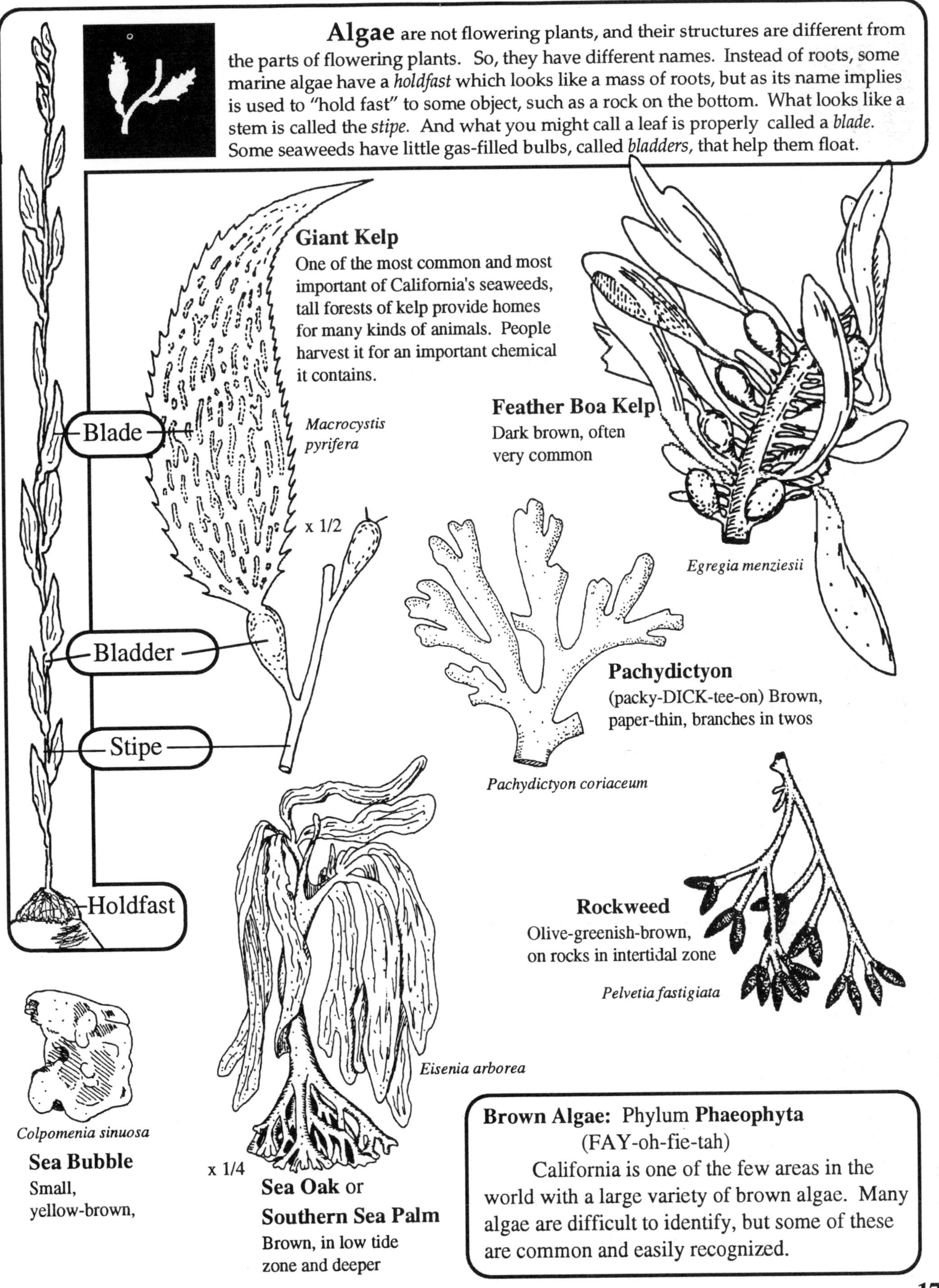

Brown Algae: Phylum **Phaeophyta** (FAY-oh-fie-tah)

California is one of the few areas in the world with a large variety of brown algae. Many algae are difficult to identify, but some of these are common and easily recognized.

Green Algae: Phylum **Chlorophyta** (KLOR-oh-fie-tah)

This group contains many fresh-water forms, but not as many *marine* (saltwater) kinds as the other algae groups. These algae are almost always some shade of green.

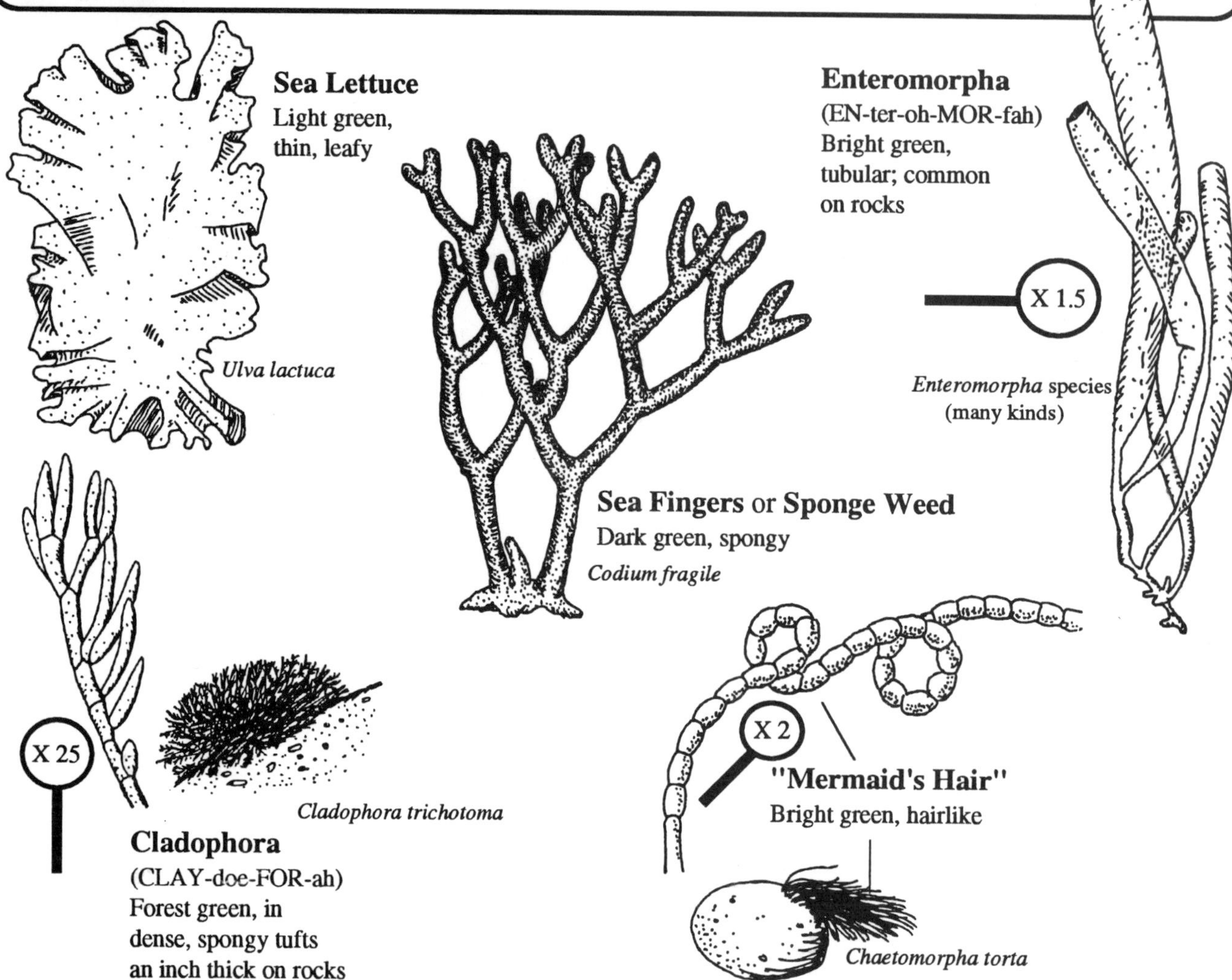

Seaweed in Your Daily Life?

California Giant Kelp is harvested by special boats, which mow it just a few feet below the surface of the water. Kelp grows back very quickly. It is harvested for *algin* (AL-jin), a chemical it contains. Algin is an *emulsifier* (ee-MULL-suh-fire), a chemical that makes two things mix together that ordinarily wouldn't, such as oil and water. Algin is used in paints, cosmetics, dairy foods and beverages as well as many other everyday products!

Make a Seaweed Collection

Sea plants can be collected, pressed between pieces of heavy *rag* or *blotter* paper and cardboard, and dried. Fully dried examples can be glued to cards and labeled. This collection can help you remember the names of seaweeds you've identified, and can be a good science project for school. You will need to get a good seaweed book and learn some specialized words, but it will be worth it.

There are some rules you'll need to follow if you want to make a seaweed collection:

Don't take live plants; take only specimens that have broken off or washed up.

Don't collect plants in parks, refuges or reserves.

Eel Grass, Surf Grass and **Sea Palm** are protected by California law and should not be collected.

Plant life is an important part of the marine food chain and environment, and should be disturbed as little as possible.

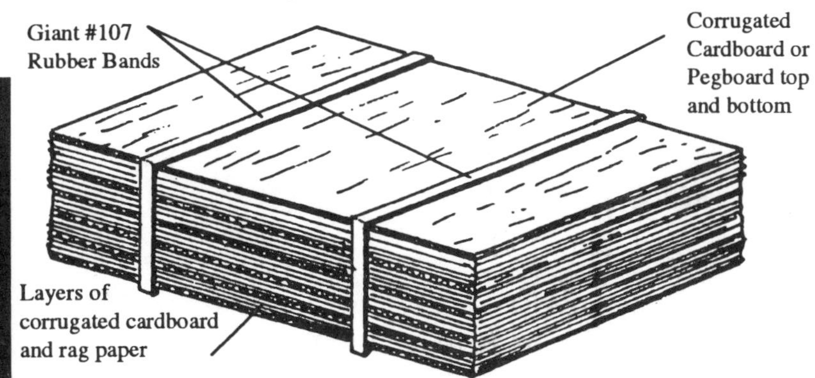

Make a Plant Press

You can make a Seaweed Press out of cardboard, rag paper and rubber bands. You can make it even more sturdy by cutting a pieces of pegboard or plywood for the top and bottom. This press can also be used for leaves or flowers of land plants. A handy size to make it is 8 by 10 inches. This is easy to carry, and your dried plants will fit on a regular 8-1/2 by 11 inch card.

Explore a HOLDFAST!

If you happen to find a fresh kelp *holdfast* on the beach, take the time to poke around in its tangled tendrils. These are ideal hiding places for many tiny animals such as crabs, worms, urchins, brittle stars, hydroids -- even baby octopus! Who knows what you might find?

(Any living things you find can be released into nearby tidepools.)

Seaweeds That Aren't Algae

Some "sea weeds" that occur in the low tide zone aren't algae. Surf Grass and Eel Grass are true *flowering plants,* resembling the grasses that grow on the land. Although they are flowering plants, the flowers and fruits are small and hidden, hard to see. But they have true stems, roots and leaves, which you *can* see!

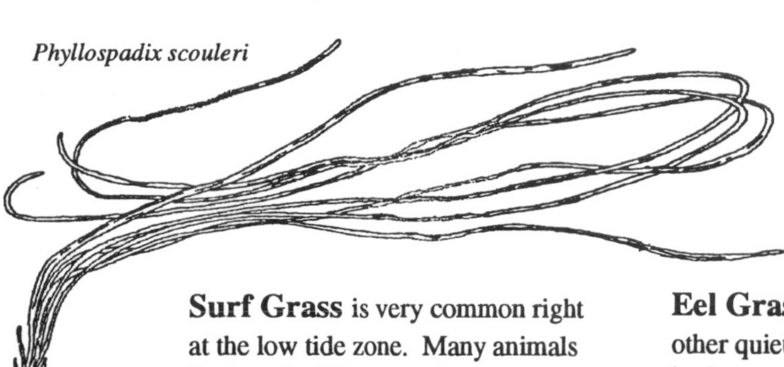

Surf Grass is very common right at the low tide zone. Many animals feed on Surf Grass, and many more hide in its long leaves and twisted roots.

Eel Grass grows in bays and other quiet water, as well as in deeper water. It isn't usually found in tidepools, but often broken pieces of it wash up into pools. Many animals use it for food and shelter.

Zostera marina

Plants That Aren't Plants!

Bryozoans (BRYE-oh-ZOH-anz) - the name means "moss animals." These colonies of small animals often resemble plants, but they aren't!

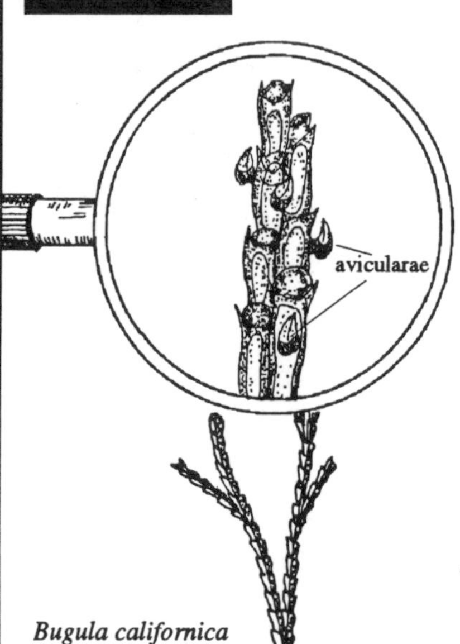

California Moss Animal

Looks like a brown or purplish clump of moss, 2 or 3 inches high. It has little, beak-like structures called *aviculare* (ay-vik-yoo-LAIR-ee) which it uses to remove unwanted material from its surface.

Jack Frost Bryozoan

A thin, encrusting form that coats almost any submerged surface with its frost-like skeletons, especially seaweed blades.

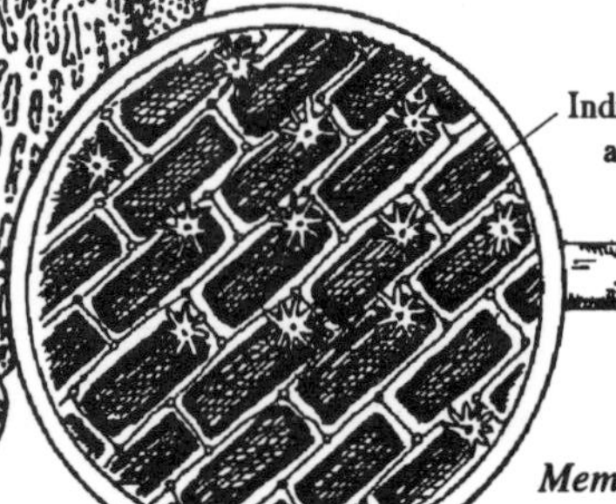

Plankton

Plankton (PLANK-tun) is the name given to microscopic, or nearly microscopic, plants and animals that float in the water or swim near its surface. Many of these are tiny organisms throughout their lives, but others are the *larval* ("baby") forms of more familiar plants and animals.

Phytoplankton (FIE-toe-plank-tun) is the name given to microscopic floating plant life (phyto means "plant," from the Greek word *phyton).*

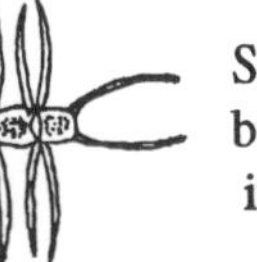

Some planktonic plants "swim" by twisting and twirling about in the water.

Diatoms (DIE-ah-tomz) are tiny plants with a silicate (glass) "shell." When the plant dies, the shell sinks to the bottom. After many, many years, a layer of *diatomaceous* (DIE-ah-tom-AY-shuss) *earth* builds up. Diatomaceous earth is used in pool filters and for other uses.

Much of the world's **oxygen** is produced by **phytoplankton!**

Dinoflagellates (die-no-FLAJ-ell-aitz) move about using a whiplike *flagellum.* One kind causes "red tides," in which sea water looks red. It also causes mussels and other animals that filter feed to be poisonous for humans to eat.

Zooplankton (ZOH-plank-tun) is the name given to the animal life in the plankton community. A great many of these creatures will grow and eventually change into more familiar tidepool animals.

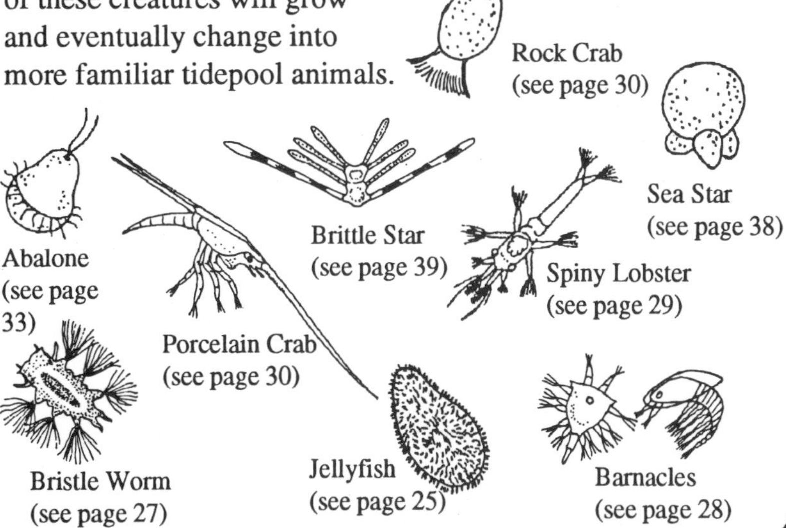

Plankton is collected in special nets pulled behind boats. It is then studied by scientists under the microscope.

Plankton is an important part of the ocean food chain - every creature depends upon it in one way or another!

Life on the Rocks

Living between a hard rock and crashing waves can be tough! Animals living in tidepools have developed many ways of protecting themselves from the surf, from conditions such as drying out or getting hot, and in many cases from each other!

In spite of all the hardships, the rocky shore can be one of the most highly populated places on earth, and nearly all the space in a tidepool is used in some way or another by animals that live there, as shown below:

Lots of animals can be found attached to the undersides of rocks or in the sand beneath rocks. Lift very gently to look, and *always* put the rock back just the way you found it. It is someone's home!

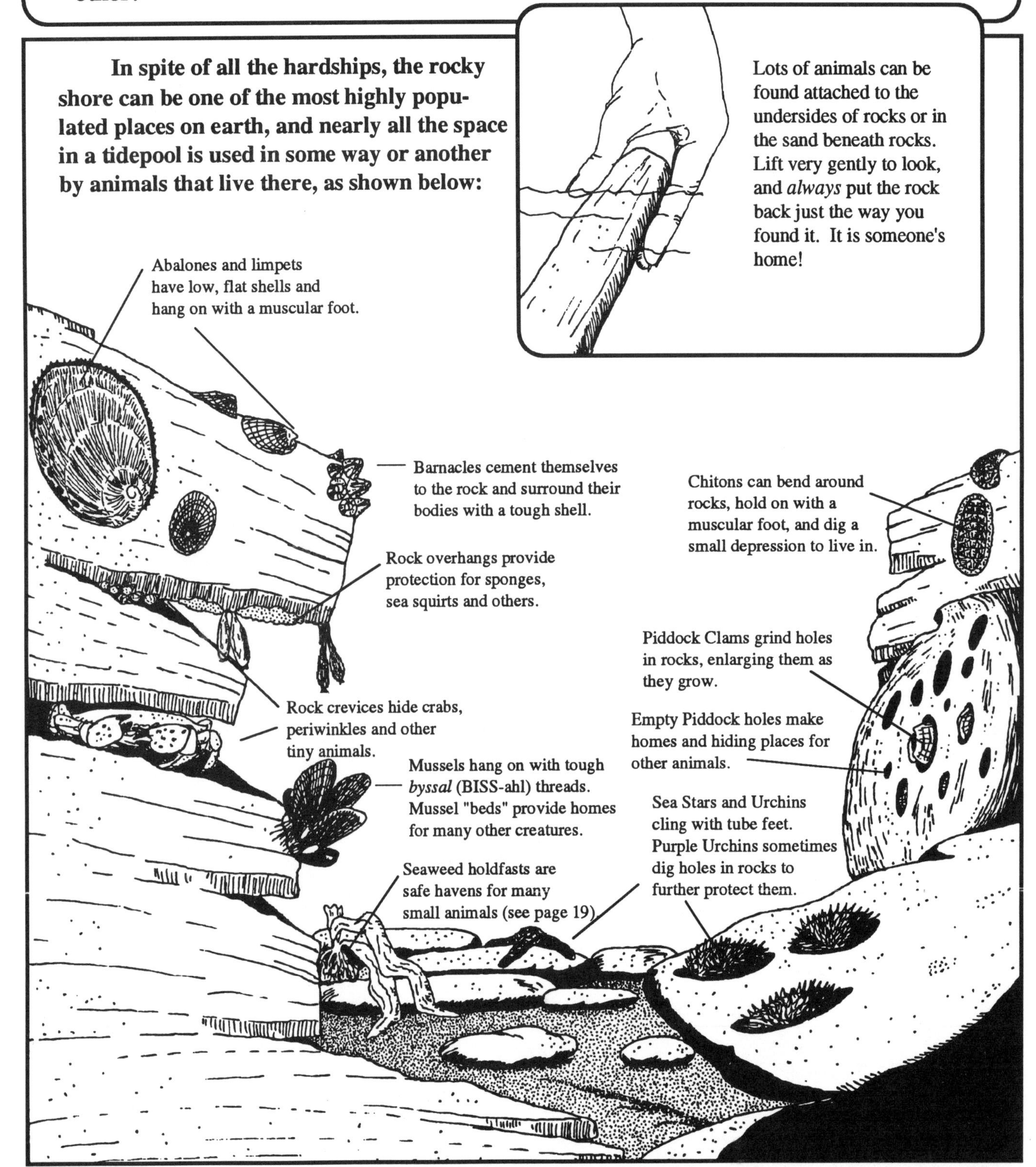

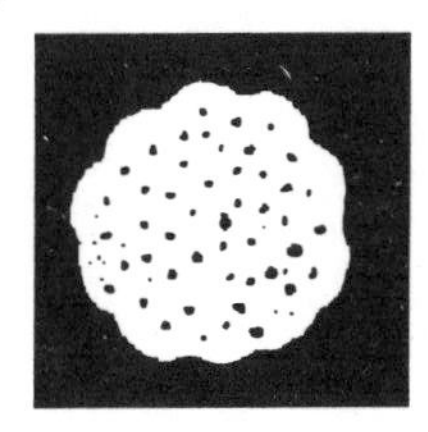

Sponges

Sponges belong to the phylum *Porifera,* which means "pore-bearing." Pores are tiny holes; most sponges are covered with tiny holes through which water passes.

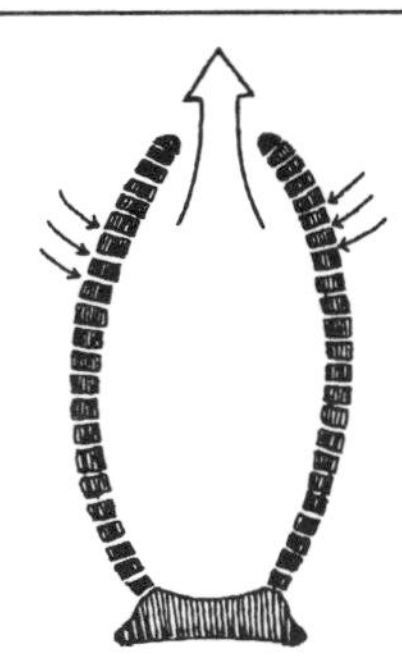

Sponges are actually colonies of cells living together. A typical sponge "body" is hollow. Water is pulled in through small holes and flows out through a larger hole.

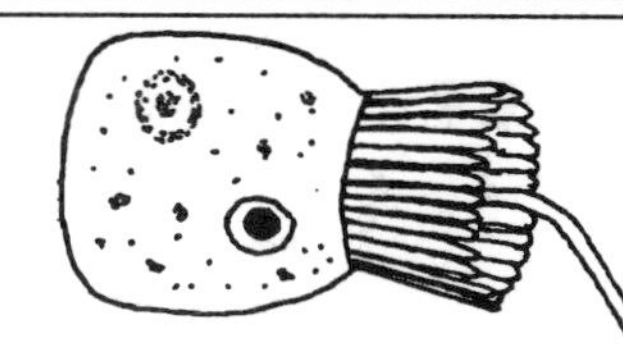

Collar Cell
enlarged many times -- you'd need a microscope to really see it

The pores of a sponge are lined with *collar cells,* which have a tiny, whip-like *flagellum* (flah-JELL-um). The rhythmic beating of thousands of these tiny flagella (plural) pulls water into the sponge body. This creates pressure inside the hollow body which forces the water out of the large opening at the top. Food particles in the water are trapped in the collar, and digested within the individual cell.

Many sponges have a kind of supporting structure (some books call it a "skeleton," but it isn't bones). This structure is made of interlocking, pointed pieces called *spicules* (SPIK-yoolz). Spicules may be made of *calcium* like seashells, or of *silica,* which we call glass. Other sponges have fibers of a protein called *spongin.* These materials give the sponge its shape, and in some cases protect it: very few animals will eat something with sharp bits of glass in it!

Most of the sponges in California tidepools are small and encrusting -- that is, they grow like a blanket or crust on rocks or other objects. One kind bores into the shells of mollusks! Its network of holes can greatly weaken the shell.

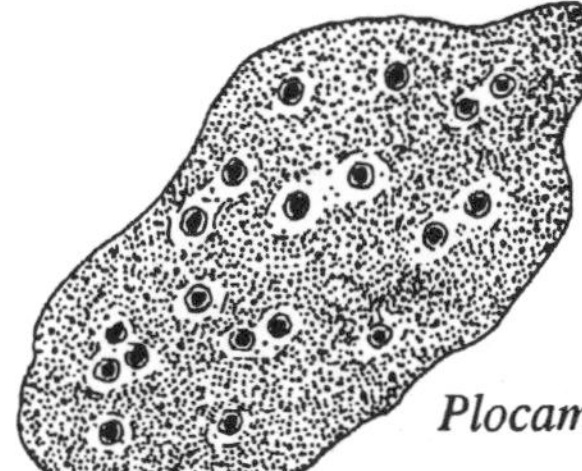

Red Encrusting Sponge
Bright red, forms mats of up to 3/4 inch thick.

Plocamia karykina

Purple Sponge
Purplish in color, with the large holes somewhat evenly spaced.

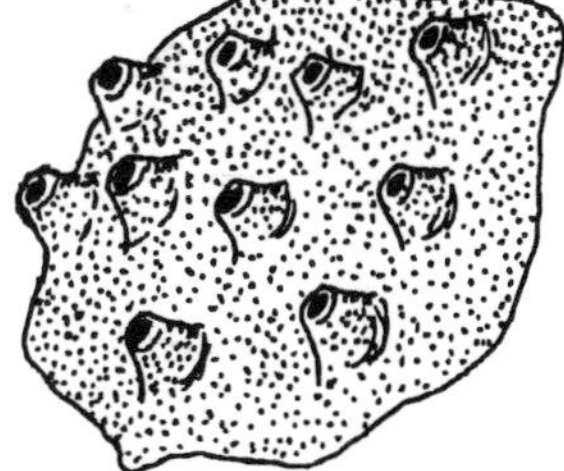

Haliclona permollis

Urn Sponge
Whitish or pale gray, hanging in clusters in rock crevices in the low tide zone.

Leucilla nuttingi

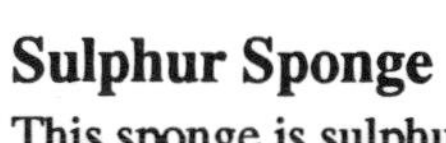

Verongia thiona

Sulphur Sponge
This sponge is sulphur yellow in color, and also has a sulphur-like "rotten egg" smell to it.

Crumb of Bread Sponge
Whitish, irregular in shape, found on the underside of rock overhangs.

Leucetta losangelensis

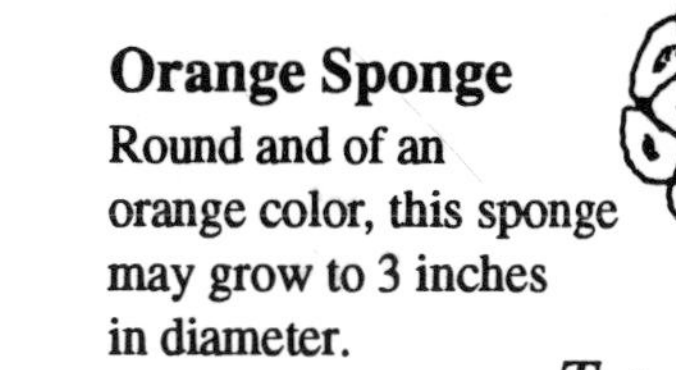

Orange Sponge
Round and of an orange color, this sponge may grow to 3 inches in diameter.

Tethya aurantia

Stinging Animals

The Cnidarians (nye-DARE-ee-anz) are a confusing group of animals. Some are colonies of animals living together, while others are individual animals; but sometimes, even scientists aren't sure which is a single animal and which is a colony! All of them have several things in common, however:

Hollow Body

In older books, Cnidarians are called *Coelenterates* (sell-ENTER-ates), which means "hollow gut." The basic Cnidarian body plan is a ring of tentacles surrounding a mouth, which opens into a hollow body.

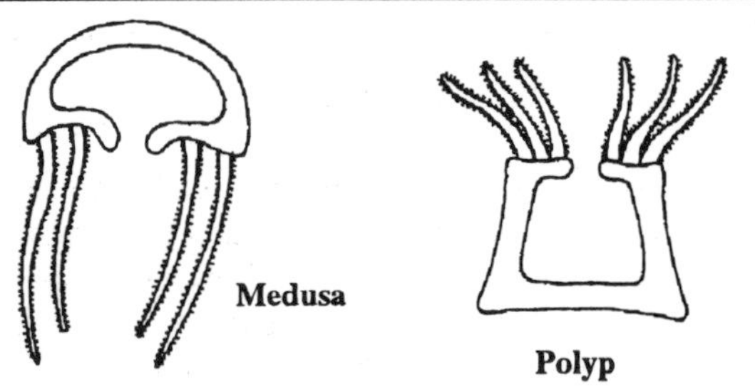

Two Body Shapes

Cnidarians generally are found in one of two body shapes: One is with the mouth facing down and the tentacles dangling. The animal is usually a swimming or floating species. This body shape is called a *medusa* (meh-DOO-sah). In the other form, the animal is usually attached to something with its mouth and tentacles facing up. This body shape is called a *polyp* (PALL-lip).

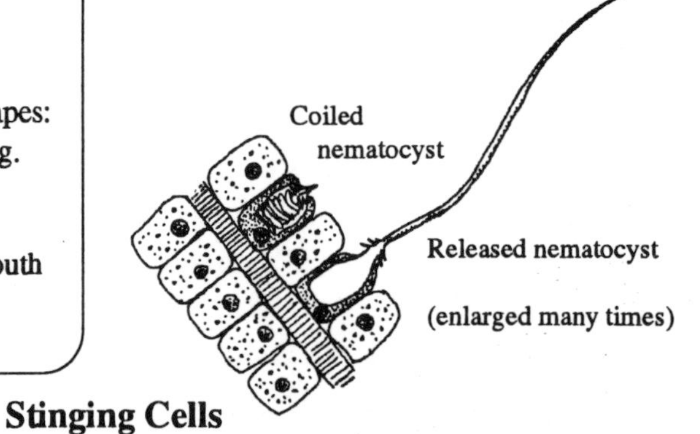

Stinging Cells

The tentacles (and sometimes other parts of the body) of Cnidarians possess specialized cells called *nematocysts* (nee-MAT-oh-sists). These are tiny capsules containing a coiled line which will burst open if disturbed. The tip of the line will penetrate prey or an enemy, and inject a bit of poison. This will stun a food animal so that it can be captured, or repel an enemy.

Hydroids

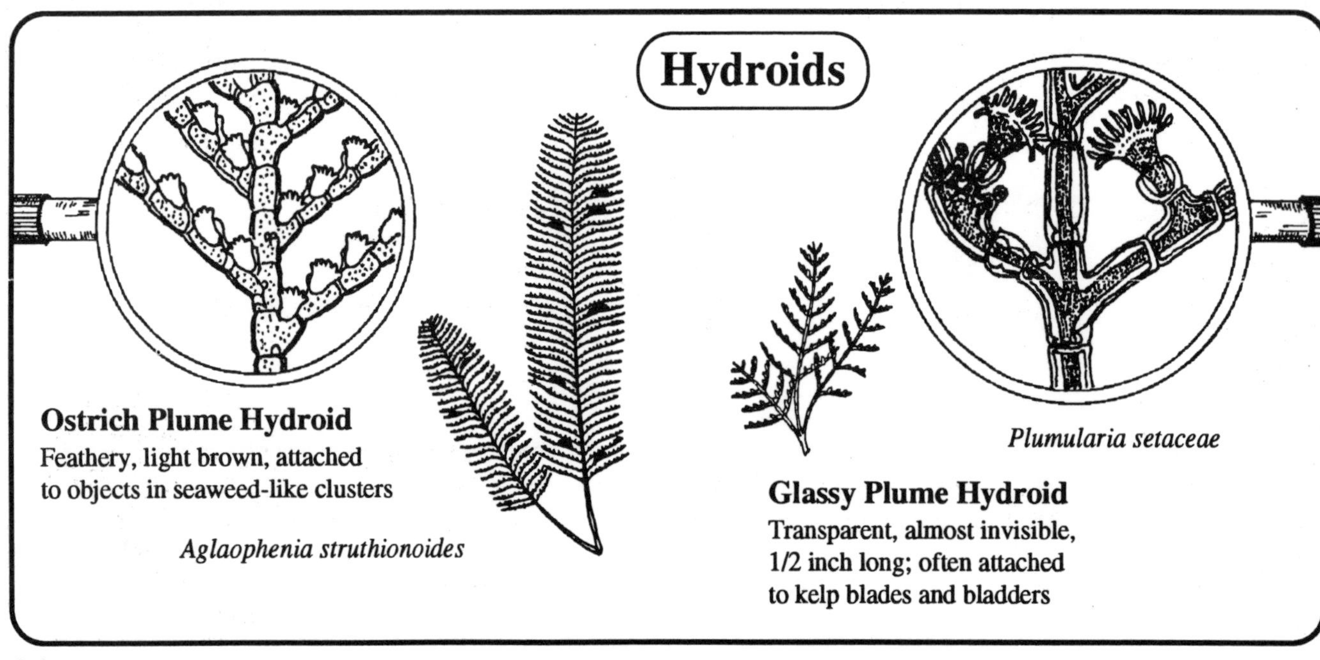

Ostrich Plume Hydroid

Feathery, light brown, attached to objects in seaweed-like clusters

Aglaophenia struthionoides

Plumularia setaceae

Glassy Plume Hydroid

Transparent, almost invisible, 1/2 inch long; often attached to kelp blades and bladders

Anemones and Relatives

Aggregate Anemone *Anthopleura elegantissima*
Colonies cover large sections of rocks at the middle tide level.

Brown Sea Fan
A deeper-water colony of animals, sometimes found washed up into tidepools. Black and orange-brown, almost coral-like.

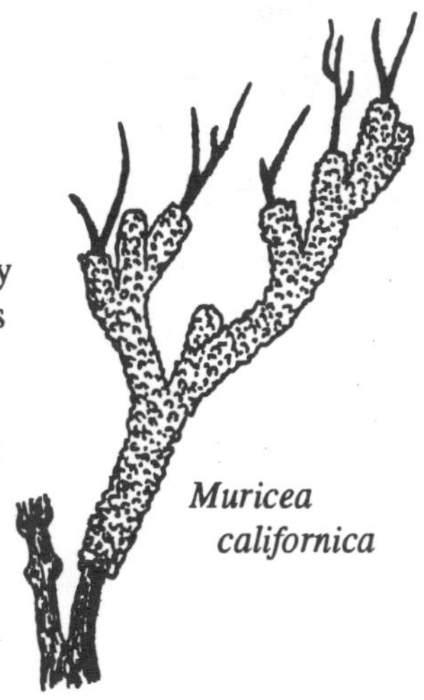

Muricea californica

The stinging cells of most of our local Cnidarians are too small to penetrate human skin. However, care must be taken not to touch a Green Anemone, for instance, and then rub your eyes or mouth. Some unsprung nematocysts may cling to your fingers and fire off in the soft, sensitive tissues around such areas. This is not dangerous, but it can be very uncomfortable!

Solitary Green Anemone
Green and purple, to 4 inches across, common in tidepools.

Anthopleura xanthogrammica

The stinging cells of some local jellyfish *can* penetrate human skin and can be quite painful. If jellyfish are in the water, you should avoid swimming there. Also, beware of jellyfish washed up on the beach -- even in jellyfish long dead the nematocysts can continue to work for a long time.

Jellyfish and Relatives

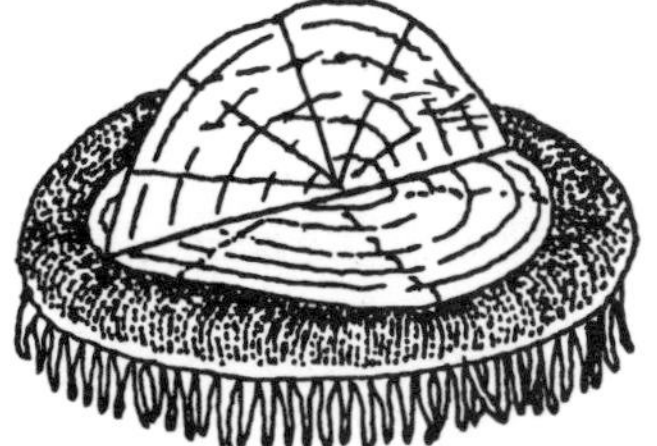

Velella velella

By-the-wind Sailor
Not a tidepool animal, but these small jellyfish are sometimes blown up onto beaches by the wind.

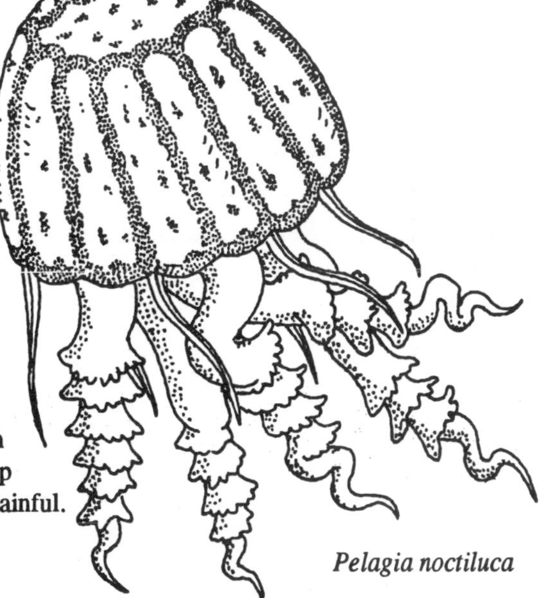

Purple-striped Jellyfish
Not a tidepool animal, but often seen in open water or washed up on the beach. Its sting can be painful.

Pelagia noctiluca

Marine Worms

There are several groups (phyla) of animals that we call "worms" because of their long, thin shape. Many of these live in and around tidepools.

Flatworms: Phylum **Platyhelminthes** (PLAT-ee-hel-MIN-theez)

Flatworms are found under rocks or ledges in tidepools. They are shy and will try to hide from you. They are very delicate, so don't try to handle them.

Common or **Brown Flatworm**
About 1/2 inch long, light brown or gray but almost *transluscent* (light passes easily through them). Under rocks in the middle and low tide zones.

Notoplana acticola

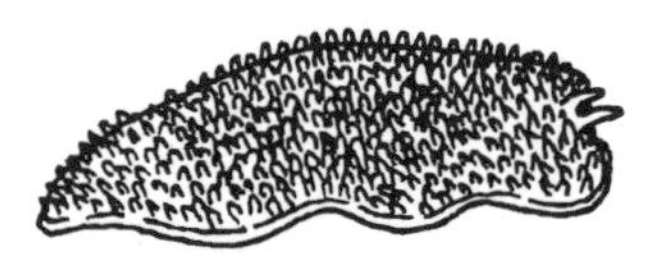

Thysanozoon sp.

Fuzzy Flatworm
About 1-1/2 inches long. The skin is covered with little structures called *papillae* (PAP-ill-ee) which look like fur. If you carefully remove this worm from its rock and place it in the water, you can watch it gracefully swim away.

Peanut Worms: Phylum **Sipunculoidea** (SIGH-punk-yoo-LOY-dee-ah)

Many Peanut Worms are about the size, shape and color of a peanut! They are found in sand around and under rocks. They strain small animal and plant bits out of the water through retractable tentacles. This is called *filter feeding.*

Sipunculus nudus

White Peanut Worm
Although 3 to 4 inches is the average length of this iridescent whitish worm, some can grow to 10 inches long!

Tan Peanut Worm
Brownish, velvety-looking, found under rocks.
Grows to 4 or more inches.

Dendrostoma pyroides

Ribbon Worms: Phylum **Nemertea** (nem-IRT-ee-ah)

Ribbon worms are *very* long and thin. They are very delicate and will break into pieces if you try to handle them. They eat using an *eversible proboscis* (pro-BOSS-iss), a "nose" that can be turned inside out and is sticky or spiked on the end.

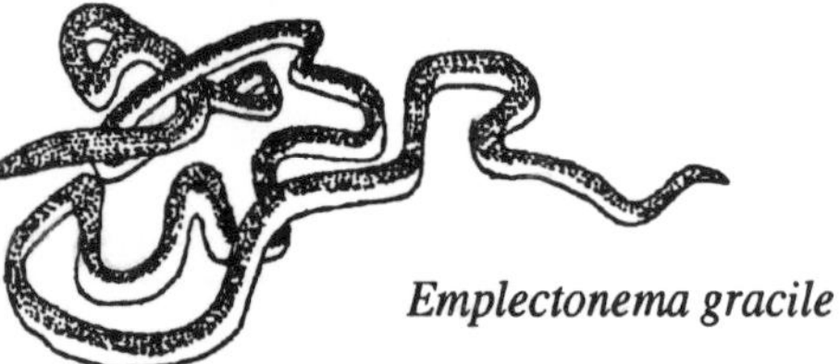

Emplectonema gracile

Green Ribbon Worm
Green above, light below, about 20 inches long. Common in mussel beds.

Banded Ribbon Worm
Olive or red-brown, banded, about 6 inches long. Found under rocks or among coralline algae in the middle tide zone.

Lineus vegetus

Segmented Worms: Phylum **Annelida** (ann-ELL-id-ah)

Segmented worms, also called *Annelids* (ANN-ell-idz), are the most commonly seen worms. Many live in freshwater and on land. You are probably familiar with the common Earthworm -- an Annelid.

Bristle Worm or Clam Worm

This is a large worm, sometimes growing to more than 3 feet long! It is olive to blue-green in color, often iridescent (shiny). It is usually found in mussel beds. Don't try to pull one out -- it will break, and it may bite you!

Nereis vexillosa

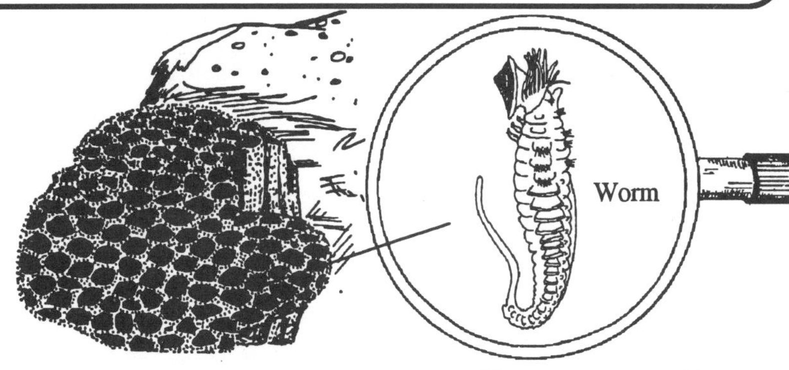

Phragmatopoma californica

Sand Castle Worm

You may never see the worms themselves, but the sandcastles they build are common on rocky beaches at low tide. They are made by cementing sand grains together with mucus. There is a worm in each tube.

Feather-duster Worm

Many species of worms can be described under this common name. They are named for their feathery-looking tentacles which they use for filter feeding and breathing. They occur in tubes attached to rocks, seaweeds or other objects, or they may be buried in sand.

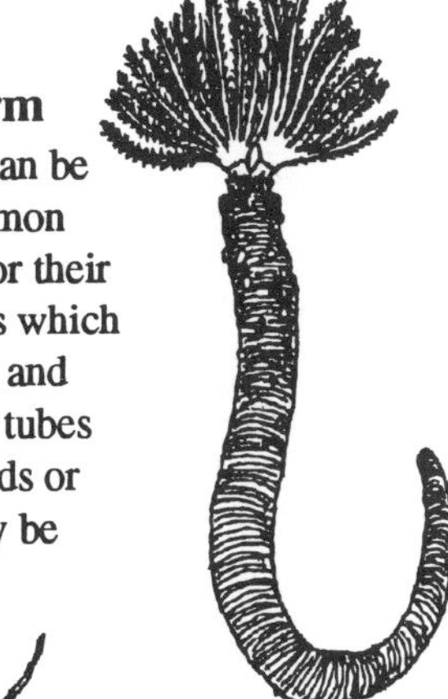

Remember that worms are very fragile, and break easily. It's best just to look at them, not to try to handle them!

Parchment Tube Worm

These worms (several types) are not tidepool residents, but are found in the mud of deeper water and the empty, parchment-like linings of their burrows often wash up into pools.

Parchment tube, often covered with bits of shell at upper end

Worm

Diopatra sp.

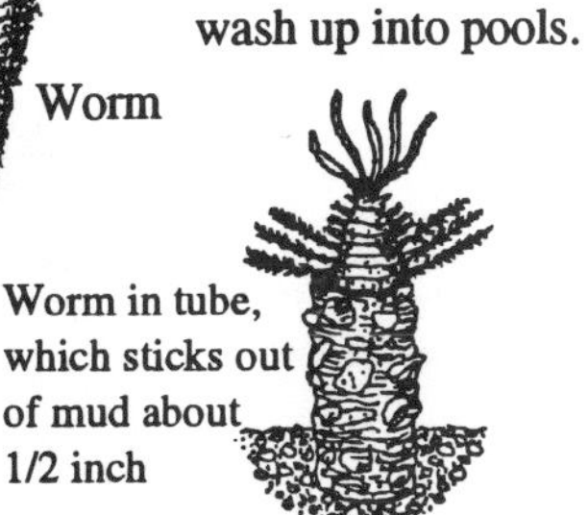

Worm in tube, which sticks out of mud about 1/2 inch

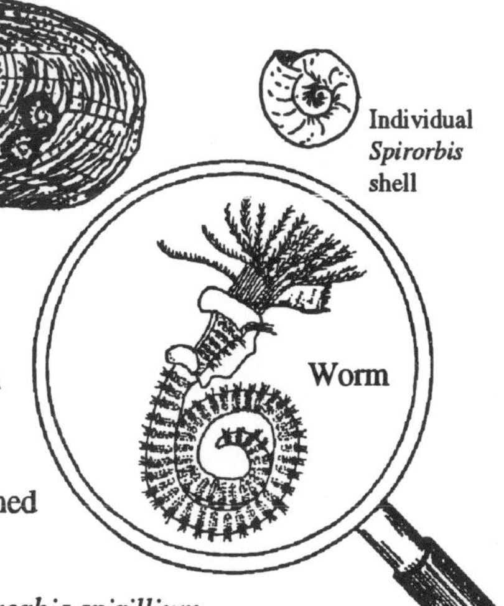

Little White Tubeworm

The tubes of this tiny worm are hard and white, usually spiraled. They are often attached to rocks or to other shells.

Spirorbis spirillium

Joint-legged Animals Phylum **Arthropoda** (ar-THROP-oh-dah)

There are more kinds of **Arthropods** (AR-throw-podz) than any other animals. In fact, one very familiar group of Arthropods -- the insects -- may be the most numerous animals on Earth!

Arthropods have a hard outer coating called the *exoskeleton.* Because their skeleton is on the outside and their muscles on the inside, they have jointed legs to move about on. Most marine Arthropods belong to the group called Crustaceans (krust-AY-shunz).

Isopods

The name isopod (EYE-so-pod) means that these animals have legs that are very much alike; none are specialized into claws. You may already know one common isopod that lives on land. It goes by the names of Sowbug, Pillbug, or Roly-poly.

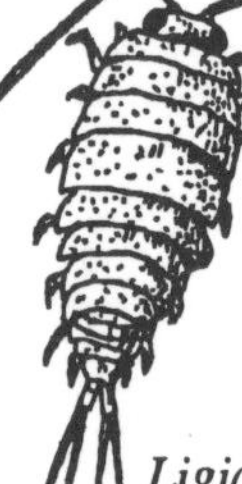

Rock Louse

Grayish, to 1-1/2 inches long, runs about rocks in the Splash Zone.

Ligia occidentalis

Idotea resecata

Kelp Isopod

Grows to 1-1/2 inches long. Color yellow-brown, matching the seaweed it lives on. Often found on kelp washed up on the beach.

Orchestoidea corniculata

Beach Hopper

Beach hoppers belong to a group called Amphipods (AM-fih-podz). They are very common on the beach at night. During the day you can find them under washed-up seaweed, which they eat.

Barnacles

Barnacles don't look much like Arthropods, but if you look at them underwater you can see them flick their jointed legs in and out while *filter feeding* and *respirating* (breathing).

Buckshot Barnacles

These are small, brownish or whitish barnacles that attach to rocks in the Splash and High Tide Zones.

Balanus glandula and *B. fissus*

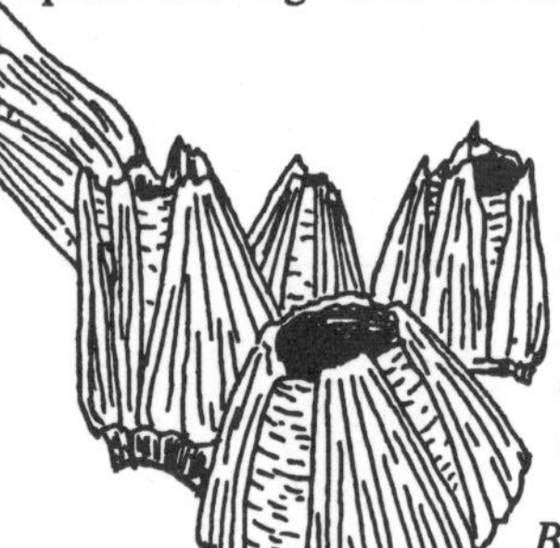

Pink and White Barnacle

A large barnacle (to 2 inches) common in the lower zones of rocky places.

Balanus tintinnabulum

Goose or **Gooseneck Barnacle**

Very common in rocky tidepools, especially among mussels.

Pollicipes polymerus

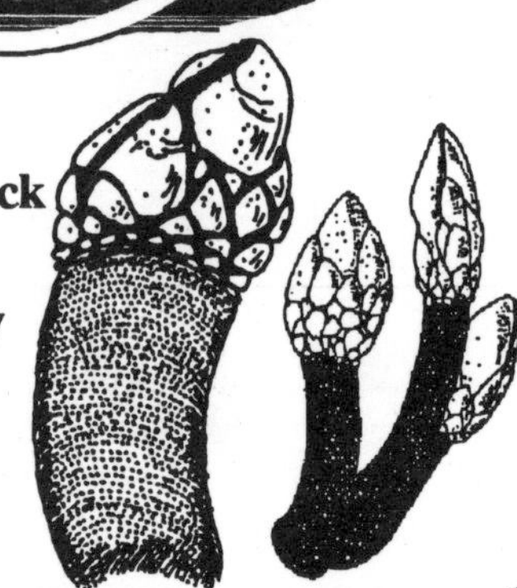

Decapods (DECK-ah-podz): Ten-footed Arthropods -- Shrimps, Lobsters, Crabs, etc.

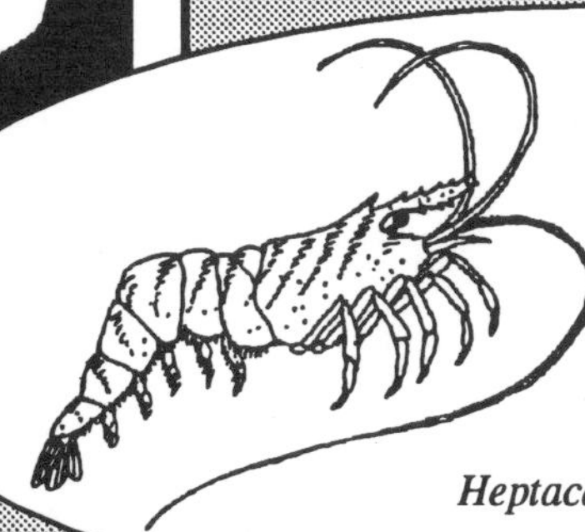

Broken Back Shrimp

Greenish-transparent with red-brown stripes. Swims by folding its tail beneath its body. Common in tidepools.

Heptacarpus pictus

Pistol or Snapping Shrimp

Only 1-3/8 inches long, this small animal can make an unusually loud sound with its specialized enlarged claw. Common along rocks at low tide, it is more often heard than seen.

Alpheus clamator

Hippolysmata californica

Striped Tidepool Shrimp

White and red striped, common in tidepools and especially easily found at night using a flashlight.

Pacific or California Spiny Lobster

These are protected by State law, so look but don't capture them.

This animal can grow to 3 feet long and up to 30 pounds in weight, but such large ones occur in deep water. Sometimes the molts (shed exoskeletons) of growing lobsters will be found on the beach, and small specimens, a few inches long, are sometimes found in tidepools.

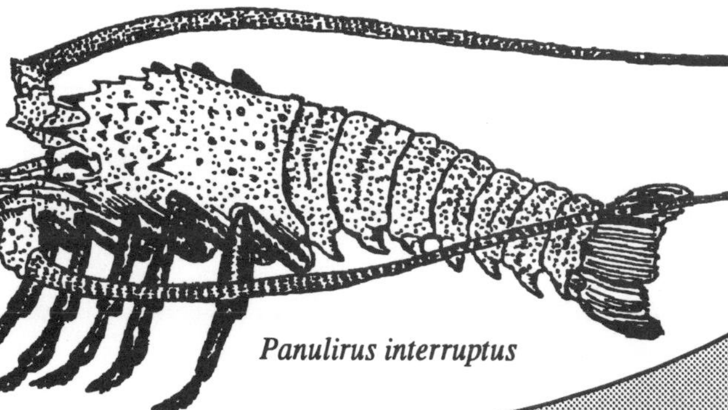

Panulirus interruptus

Blue-clawed Hermit Crab

Hermit Crabs have hard shells over the fore part of their bodies and claws, but not over their soft, coiled abdomens; so they slip into empty mollusk shells to protect their backsides! As the crab grows, it needs to find larger and larger shells to move into. Hermit Crabs are scavengers and are very common in tidepools.

Pagurus samuelis

More Decapods!

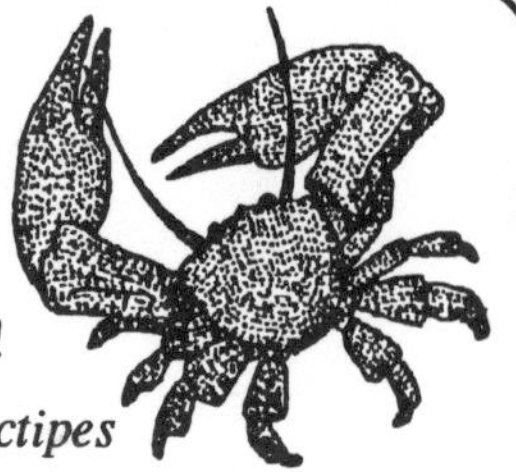

Flat Porcelain Crab
Dark gray or green, usually found under rocks. Handle carefully or not at all -- it easily sheds legs and claws!

Petrolisthes cinctipes

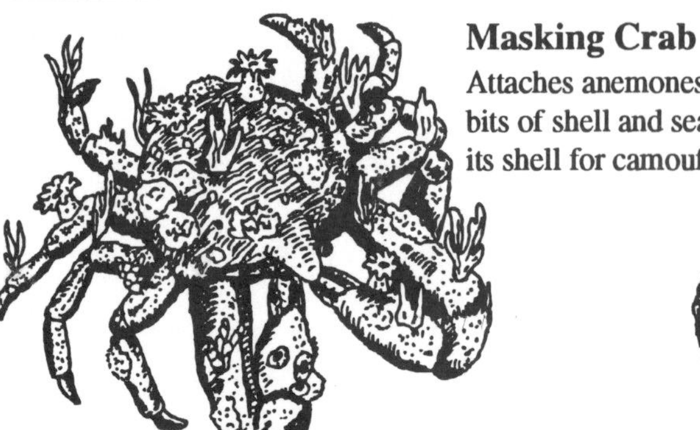

Masking Crab
Attaches anemones, sponges, bits of shell and seaweed to its shell for camouflage.

Loxorhynchus crispatus

Kelp Crab
Small crabs are found in tidepools, larger ones live in deeper water in the kelp beds. Usually the same color as the kelp it is on.

Pugettia producta

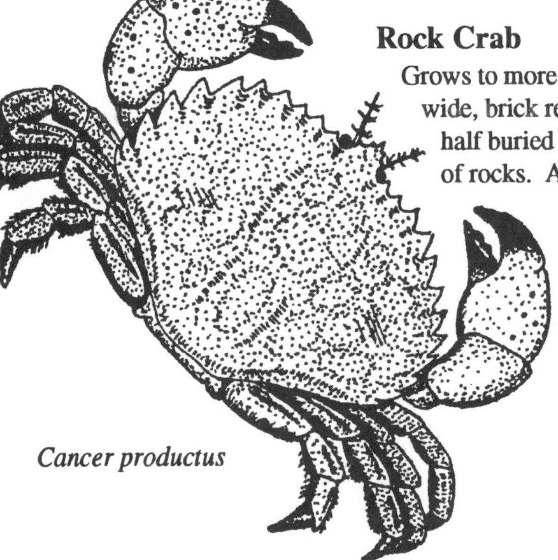

Rock Crab
Grows to more than 6 inches wide, brick red, usually found half buried in sand at the base of rocks. Another place you might see it is in seafood markets!

Cancer productus

Striped Shore Crab
A very common crab, the one you are most likely to see on rocks from the Splash to the Middle Tide Zones.

Pachygrapsus crassipes

Beware of pincers! The large front claws of *Cancer* crabs and others can give you a very hard pinch -- they might even break the skin! It's best not to try to handle these animals.

Boy Crab or Girl Crab?

If you find a crab molt on the beach and look at its underside, you can tell if it was from a male or a female crab.

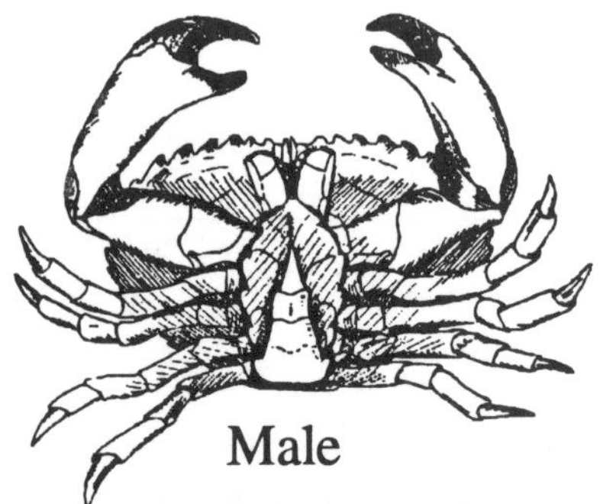

Male

It's easy to see the tail on a lobster or a shrimp. But crabs keep their tails tucked up under. On the male crab the tail is slender and pointed, while on the female it is broad and rounded.

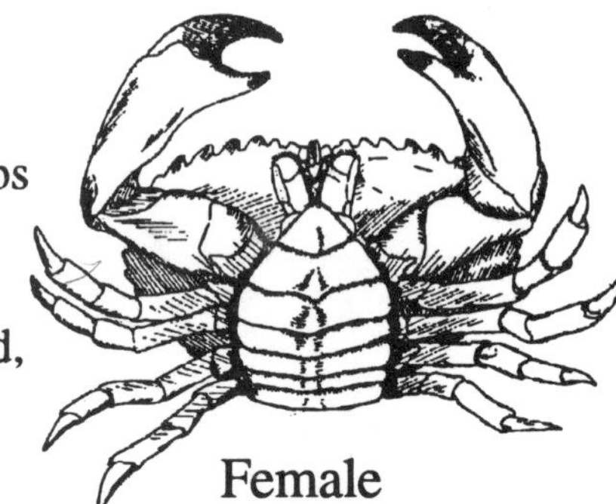

Female

The A-MAZE-ing Hermit Crab

The Hermit Crab has no hard shell of its own to cover its soft *abdomen* (the back part of its body), so it moves into the empty shells of mollusks for protection. As it grows, the Hermit Crab needs to find bigger and bigger shells.

Help this growing Hermit Crab find his way through the maze to a new shell!

Mollusks Phylum **Mollusca** (maw-LUS-kah)

Mollusks (MAW-lusks) are soft-bodied animals. Most mollusks live in shells, which are made of proteins and the mineral calcium, and are produced by a part of the body called the *mantle*. You may already know some mollusks from your garden -- slugs and snails!

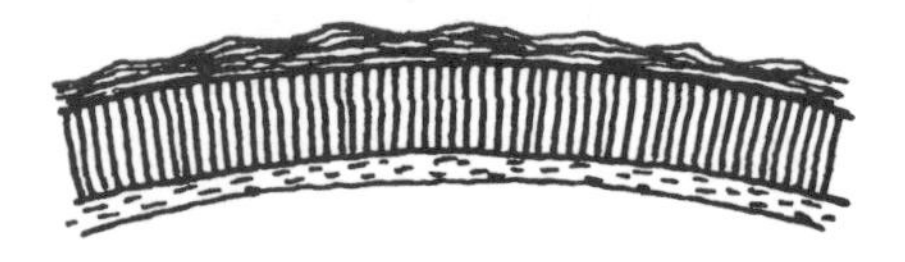

Most mollusk shells consist of three layers. The outer layer is a tough coating; the middle layer is thick and limey (chalky); and the inner layer, next to the animal's body, is smooth and sometimes shiny.

Mollusks with shells are divided into two major groups:

Those that have a single, usually spiraled shell are called *Gastropods* (GAS-troe-podz). They are also sometimes called *Univalves* (YOO-nee-valvz).

Those that have two shells which close together are called *Pelecypods* (pell-ESSY-podz). These are sometimes called *Bivalves* (BYE-valvz).

Gastropods (Univalves)

Limpets are flattened snails that live on rocks from the Splash Zone down into pools.

Rough Limpet
Collisella scabra

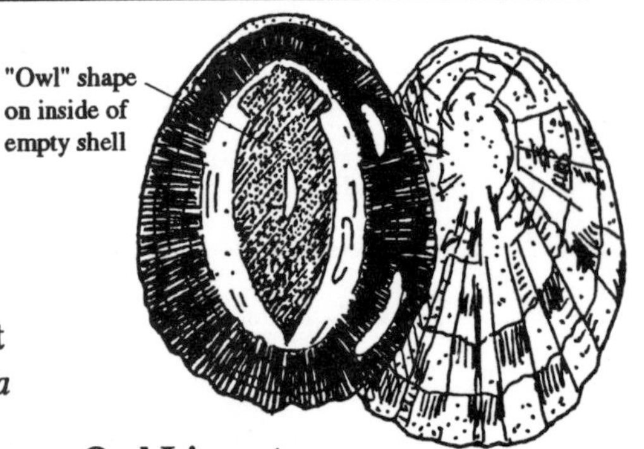

Owl Limpet
Lottia gigantea

Shield Limpet
Collisella pelta

Kelp Limpet
Notoacmea insessa
Brown, usually on Feather Boa Kelp

Slipper Shells resemble limpets, but you can see a little "shelf" inside an empty shell.

Onyx Slipper
Crepidula onyx

Often stacked on top of one another

Turban Slipper
Crepidula adunca
Found on Black Turban Snail shells

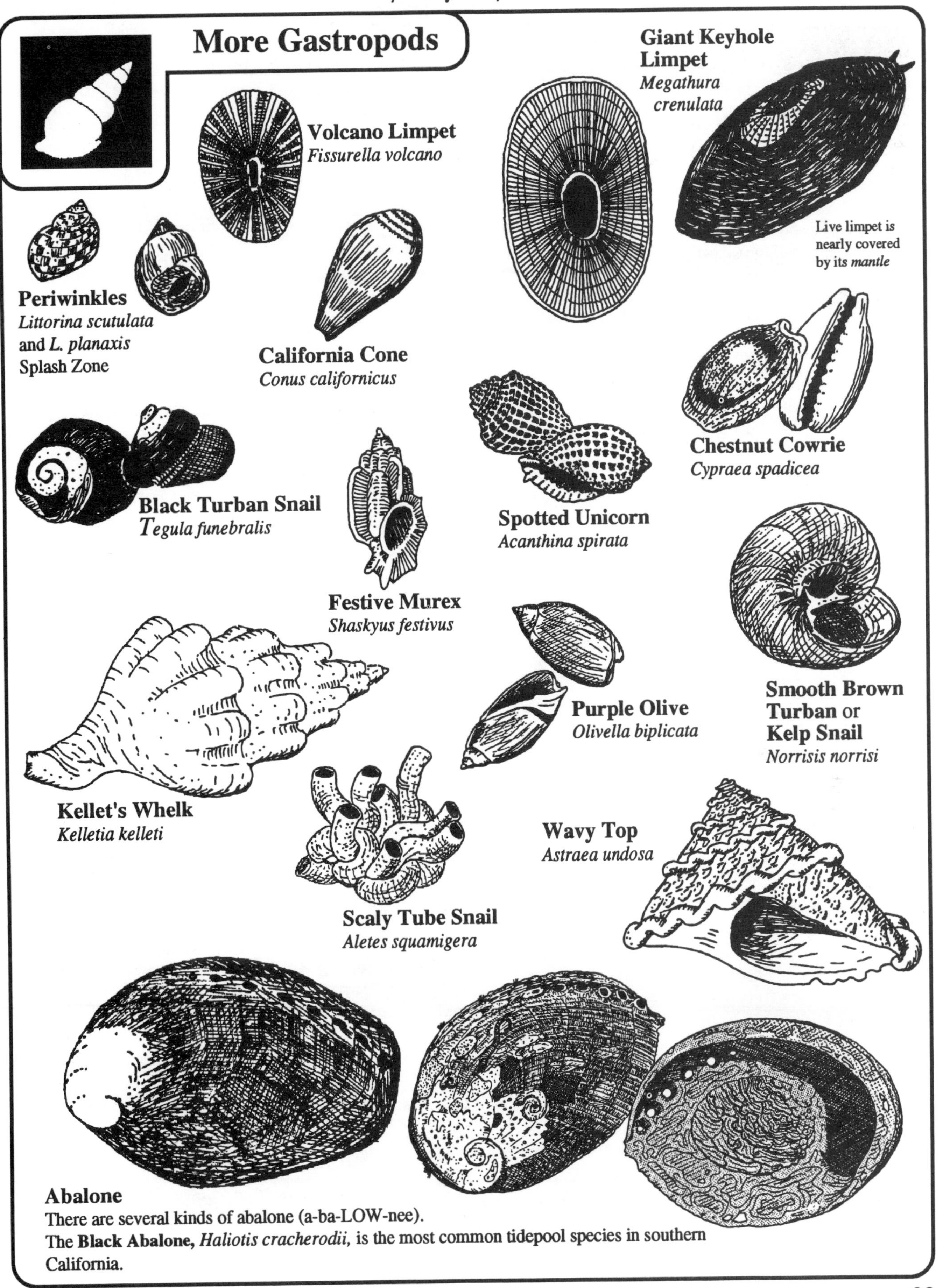

Abalone

There are several kinds of abalone (a-ba-LOW-nee).

The **Black Abalone,** *Haliotis cracherodii,* is the most common tidepool species in southern California.

Gastropods Without Shells

Whoever heard of a snail without a shell? Chances are, YOU have! You've probably seen a garden slug. That's basically a snail without a shell. In the tidepools, there are several types of gastropods that have no shells, or very small shells that are hidden within their mantle. These are commonly called "Sea Slugs."

Nudibranchs (NOO-dee-branks) means "naked gills;" these are sea slugs that have their respiratory (breathing) structures (called *cerata* [sair-AH-tah]) exposed.

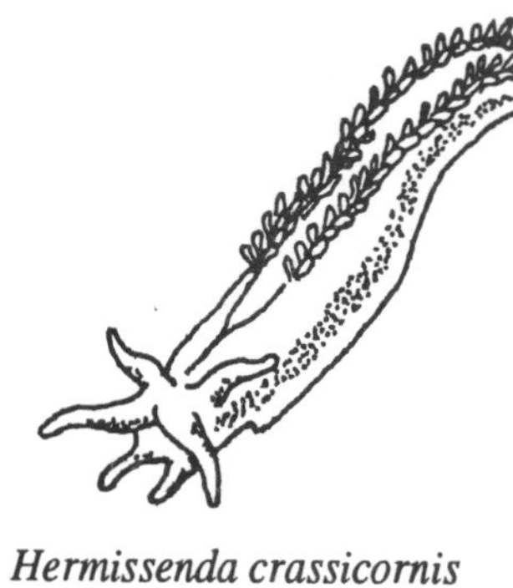

Hermissenda crassicornis

Hermissenda

(her-miss-END-ah)

Grows to more than 3 inches long; body gray-white with an orange stripe over the back, and iridescent blue stripes along the edges. Eats cnidarians, and also other nudibranchs!

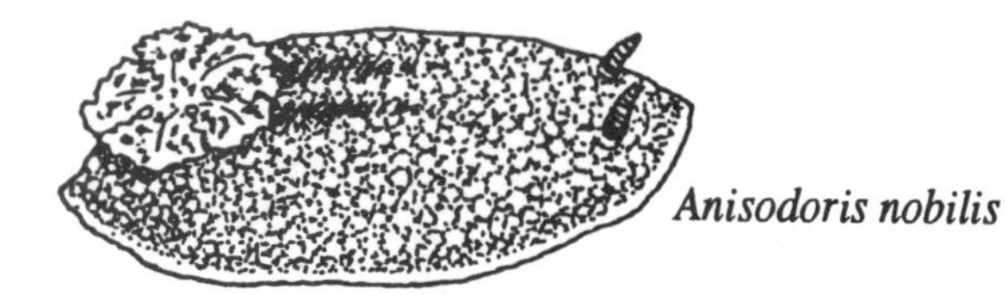

Anisodoris nobilis

Lemon Nudibranch

Coarse and bumpy-looking, usually some shade of yellow with brown spots. Gill plumes all at one end, pearly white. Eats sponges.

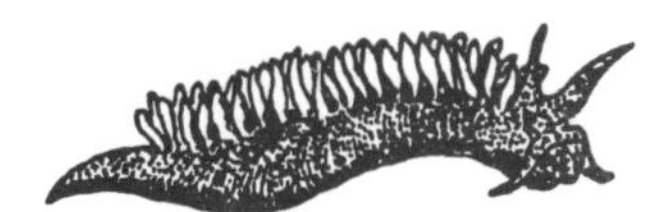

Flabellina iodinea

"Spanish Shawl"

Color bright purple, with orange *cerata* (gill structures). "Swims" with graceful undulating motions.

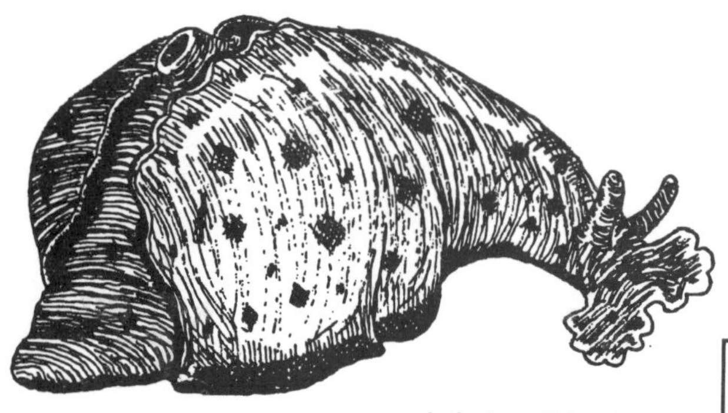

Aplysia californica

Sea Hare

Largest of local "sea slugs," this creature grows to more than a foot long! It is spotted, brown or purplish, and is common at minus tides in the beds of surf grass and kelp upon which it feeds. It can be handled carefully, but if it gets bothered it will emit a deep purple "ink" to try to protect itself.

The Sea Hare is an *opisthobranch* (oh-PIST-oh-brank), which means "covered gills;" Its respiratory structures are not exposed, but are covered by its mantle.

Sea Hares lay stringy, tangled, yellowish egg masses, some as large as a grapefruit. Such masses may contain 86 million eggs! The young are free-swimming, and become part of the *plankton* eaten by fish and other animals. Otherwise, we'd be knee-deep in Sea Hares!

Pelecypods (Bivalves)

Look at a Clam Shell:

Look on the inside of a clam shell. You can see some of the features of a typical pelecypod (pell-ESSY-pod).

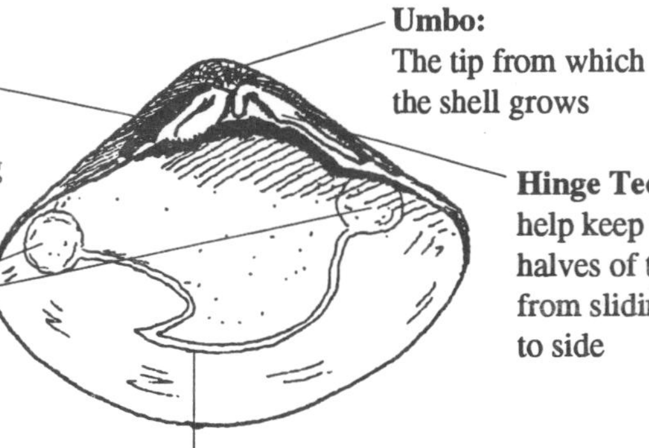

Many pelecypods live in sand or mud, using their muscular foot to burrow and extending their siphons (SIGH-funz) to the mud's surface to obtain oxygen and food.

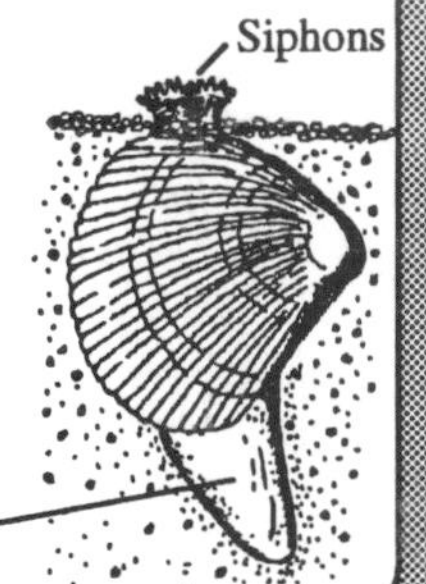

Chama or **Jewel Box**
The bottom shell is attached to rocks.
Chama pellucida

Basket Cockle
Clinocardium nuttallii

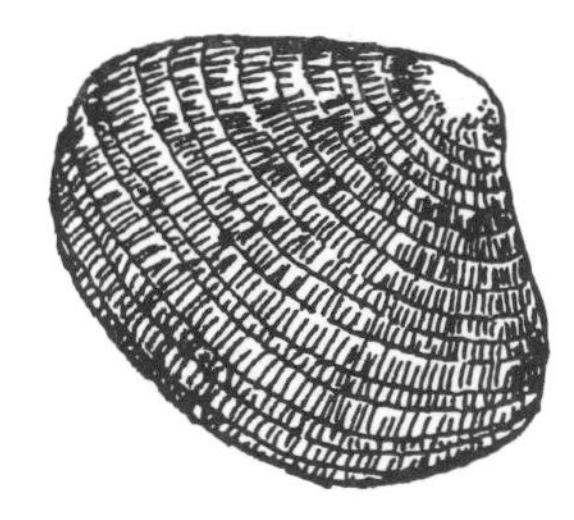

Common Littleneck
Protothaca staminea

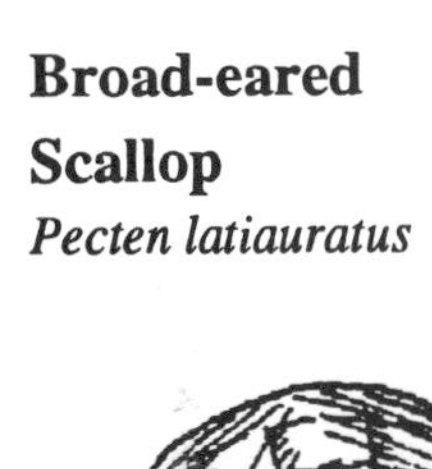

Broad-eared Scallop
Pecten latiauratus

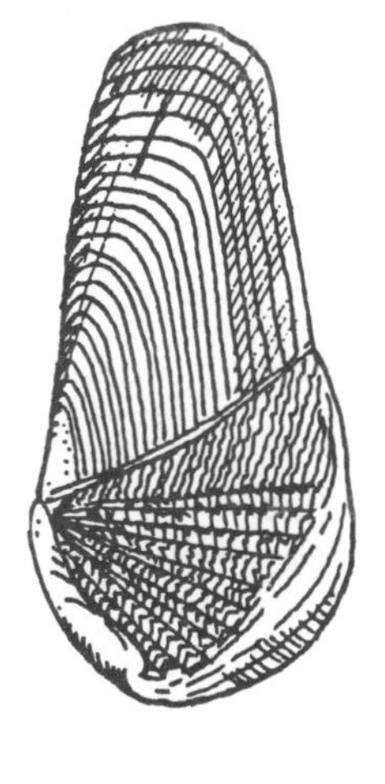

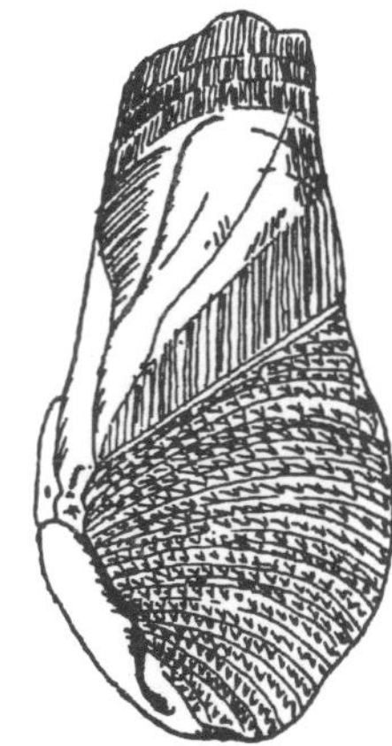

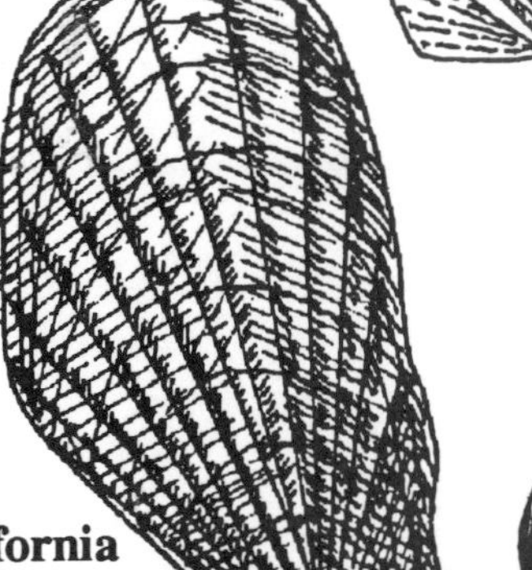

California Mussel
Mytilus californianus

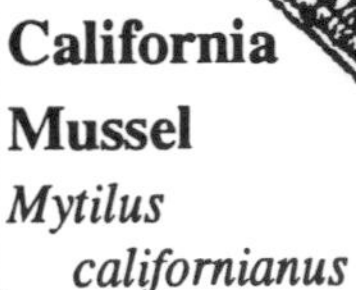

Bay Mussel
Smooth, blue, found in quiet waters
Mytilus edulis

Piddock Clams
Use the enlarged, rough end of the shell to bore holes into soft rock.
Zirfaea pilsbryi and other species

More Mollusks

In addition to the two main groups of mollusks, there are several smaller groups. Two of these groups are frequently seen in southern California.

Chitons

Members of the **Amphineura** (am-fin-YOO-rah) are commonly called **Chitons** (KYE-tunz) or Sea Cradles. Chitons have an 8-part shell. Sometimes "butterfly shells" are found in tidepools; these are segments of a chiton shell.

"Butterfly Shell"

Nuttallina californica

Rough Chiton
Common on rocky shores from high to mid tide zones.

Mopalia ciliata

Mossy Chiton
Common at mid to low tide zones. Branching bristles on the mantle and dark green color make it look "mossy."

Cryptochiton stelleri

Gumboot Chiton
The world's largest chiton, grows to slightly more than a foot long; leathery, brick-red mantle covers shell. Mid to low tide zones.

Cephalopods

The **Cephalopod** (SEF-al-oh-pod) group, which means "head-foot," includes the mollusks we call squids and octopi. It is hard to think of these animals as mollusks, but the internal remnant of a shell shows their relationship to the rest of this group.

Two-spotted Octopus
This is the most common tidepool octopus in southern California. It will be found under rocks or in small caves. It will not hurt you, but it is very delicate and should not be handled. If you can coax one into your bucket with a net, you will be able to get a closer look at it. Octopi (which is what you call more than one octopus, although you can also call them octopuses or octopods!) eat crabs, clams, mussels, scallops, worms and fish.

Octopus bimaculatus

Make a Shell Collection

Making a shell collection can be fun and educational. But there are lots of things to consider before you make one:

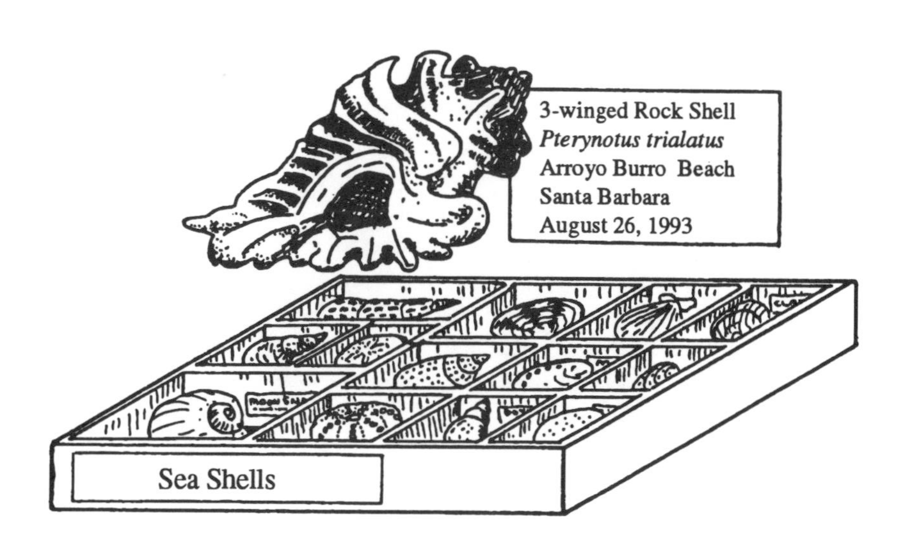

Marine life is protected by California law and in all parks and preserves. In some of these you may not be allowed to collect shells. Be aware of the laws in the places you visit, and obey them.

Empty shells are an important part of the marine ecosystem. Hermit crabs and other animals live in them, and the calcium they are made of is recycled for use by others. Shells belong where they are unless needed for a *useful* collection.

A shell collection is no good if it is not used. Shells stored away in boxes or cabinets are not useful. If you don't plan to keep the shells and display them, it's best not to collect them.

A good collection consists of one fine shell of each type. Don't collect lots of the same thing.

Never collect shells with live animals in them (be especially aware of Hermit Crabs which may pull in and be hard to see). If you don't find an empty shell of the kind you want, wait until another time. Eventually you will find one, and the hunt for a certain shell can be a special adventure.

Don't collect one shell per person if you are making a classroom collection; collect just one of each type of shell.

Make an attractive case for your shells, and make labels telling what each shell is, where it was found, and when. You will need a good shell book for your area to help identify them.

Spiny-skinned Animals Phylum Echinodermata

(ee-KINE-oh-der-MAH-tah)

The **Echinoderms** (ee-KINE-oh-dermz) have bodies divided into fives (not always obvious), and move about using **tube feet** which are operated by a series of bulbs and tubes that carry water throughout the body.

Sea Stars

Pisaster ochraceus

Common Sea Star
Sometimes called Ochre Sea Star; ochre is a yellow-orange color, but while this animal *may* be ochre, it may also be brown or purple!

Knobby Sea Star
Brown with white spines, which have a blue ring at the base.

Pisaster giganteus

Bat Star
Comes in a variety of colors! Its tube feet don't attach to things, so this star can be *very gently* handled.

Patiria miniata

Rainbow Star
Green, red and pink - very pretty, but very slimy and breaks easily, so look but don't handle.

Astrometis sertulifera

Six-rayed Star
(sea star "arms" are properly called *rays)*

Leptasterias haxactis

Sunflower Star
Pycnopodia helianthoides

Variable Star
Linckia columbiae

Leather Star
Dermasterias imbricata

Pisaster brevispinus

Short-spined Star
Pink, sometimes very big!

Brittle Stars

Ophioderma panamensis

Serpent Star
Brown; look under rocks

Brittle stars are thin and fragile, and easily break into many pieces. It's best not to try to handle them!

Spiny Brittle Star
Ophiothrix spiculata

Ophionereis annulatus

Banded Brittle Star
Brown and gray, found under rocks or in kelp holdfasts. Very brittle!

Sea Urchins

Purple Sea Urchin *Strongylocentrotus purpuratus*
Sometimes very common in rocky areas. The empty shell is called a "test."

Strongylocentrotus franciscanus

Red Sea Urchin
Long spines, deep red in color; found in Low Tide Zone

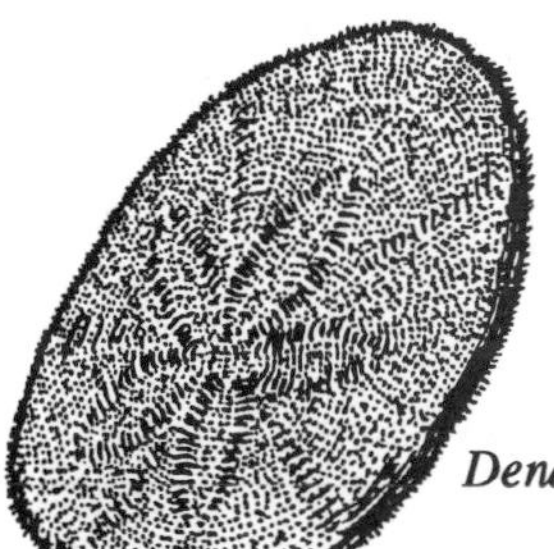

Sand Dollar
Sand dollars are flat sea urchins with tiny, velvety spines. They live in deep water, in the sand. Sometimes their empty "tests" wash up on the beach.

Dendraster excentricus

Sea Urchins have a five-piece "jaw" called "Aristotle's Lantern." The spines of some urchins are poisonous, but NOT the ones in southern California!

Sea Cucumber

Sea cucumbers are the strangest echinoderms of all. They don't always look spiny, but they often have lots of bumps. And look for the tube feet!

Stichopus parvimensis

Common Sea Cucumber
Orange-brown, up to almost a foot long! Must be handled very carefully or it may *eviscerate* (ee-VIS-sir-ate) -- literally throw all its "guts" out! (It can grow a new set in about 2 months).

Going Bananas with Echinoderms!

We mentioned that Echinoderms are animals with their bodies divided into fives, although it is not always obvious. You can get an idea of Echinoderm structure by using a banana peel. Take a fresh banana and *carefully* split the peel into five fairly equal strips. Don't separate them at the end, leave them attached to the tip.

Lay your banana peel out flat, and you have the basic design of a Sea Star!

Now, pull up the tips of your "Sea Star's" arms and hold them together at the top; imagine the empty spaces filled in, and the whole little globe covered with spines. What do you have? a Sea Urchin!

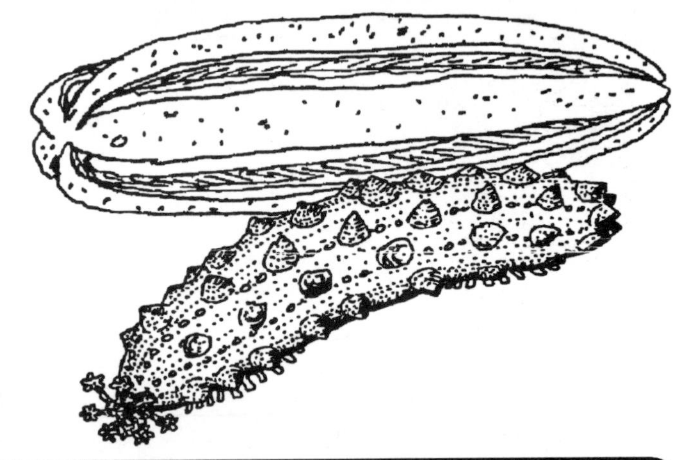

Turn your banana urchin on its side and stretch it out. You now have a banana peel Sea Cucumber! Real Sea Cucumbers have five long strips of muscles inside their bodies. These allow them to stretch out long and thin or pull up short and fat!

Don't forget to eat your banana -- its good for you! And put the peel in a trash can, compost heap or some other proper place. Don't leave it on the beach or in the tidepools!

Exploring the Stars!

Mussels vs. Muscles

Some Sea Stars eat clams and mussels. They use their hundreds of tiny tube feet to constantly pull at the two halves of their victim's shell. Clams only have two muscles (mussels have one) to hold their shell closed; eventually they get tired and their shell opens -- just a little! But that's all the Sea Star needs: It turns its stomach inside-out, inserts it into the open crack, and digests the clam inside its own shell!

These tube feet are what the Sea Star also uses to hold on to the rocks against pounding waves. That's why you shouldn't try to pick up this kind of Sea Star. You'll end up tearing the tube feet and exposing the Sea Star to danger if it can't hang on.

Touch the Stars!

You shouldn't pick up a Common Sea Star, but here's something you *can* do. Lay the top of your arm gently on top of a Sea Star and leave it for a few seconds. When you pull away, it will seem like the Sea Star stuck to the hair on your arm! How did this happen? Get out your magnifying glass to find out!

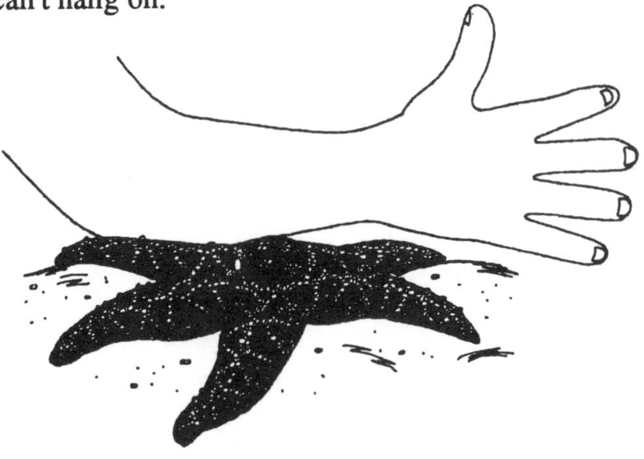

Pedicellariae

Tube Feet

Under the magnifying glass, you can see that the Sea Star has tiny, pincer-like structures on it. These are called *pedicellariae* (ped-iss-sell-AIR-ee-ay). They keep the Sea Star clean. That's what grabbed you! And that's why you don't see barnacles or kelp growing on Sea Stars.

Strange Cousins

Phylum **Chordata,** subphylum **Tunicata**

Tunicates (TOO-nuh-kaytz) don't look like they are related to animals with backbones, but they are. They possess a *notochord* (NO-tow-cord), which is like a rudimentary backbone. They are sometimes called *Sea Squirts.*

Compound Ascidian

(as-SID-ee-an)

Forms thin crusts or rubbery blobs around objects. Color varies from red to purplish. Each "petal" of the flower-like structures is an individual, pinhead-sized animal..

Botrylloides diegensis

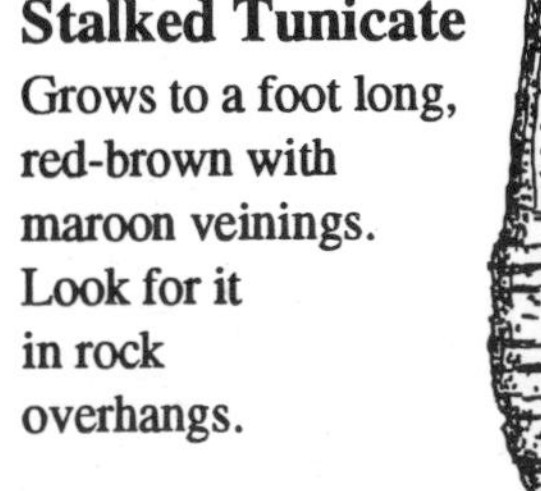

Sea Pork

Orange and rubbery, resembles a piece of salt pork. Frequently washes up into pools.

Amaroucium californica

Stalked Tunicate

Grows to a foot long, red-brown with maroon veinings. Look for it in rock overhangs.

Styla montereyensis

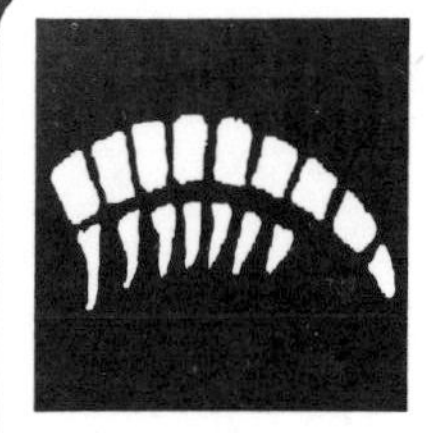

Tidepool Fishes

Phylum **Chordata,** subphylum **Vertebrata**

Small fish can often be found in tidepools. Some of them are the young of kinds that will live in deeper water when they grow up. Others are small kinds that may spend their whole lives in tidepools. There are hundreds of kinds of fish that you may see. A good tidepool fish book and a lot of patience will reward you with lots of enjoyable fish-watching.

Sharks and Rays

Sharks and rays have skeletons made of cartilage, the soft, flexible material found at the ends of your bones and in your nose. Most sharks are harmless, and several kinds may be found in tidepools when they are very young.

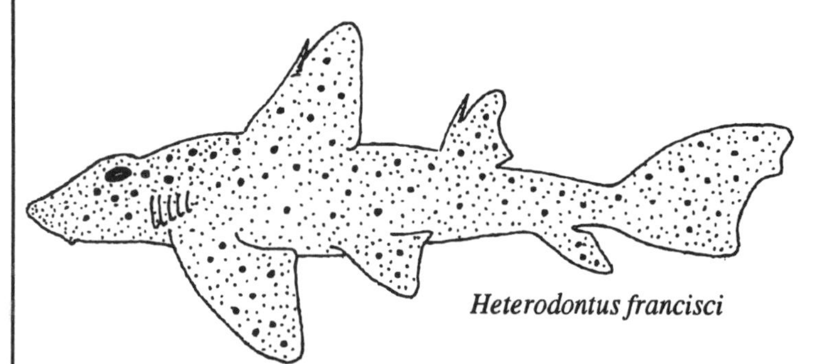

Horn Shark
(so called because of the spines in its upper fins)

Heterodontus francisci

Horn Shark Egg Case

Catch fish in your net, watch them for a few minutes in your pail; but be *very careful* and be sure to put them backwhere you found them!

Bony Fishes

Most of the world's fishes have skeletons made of true bone.

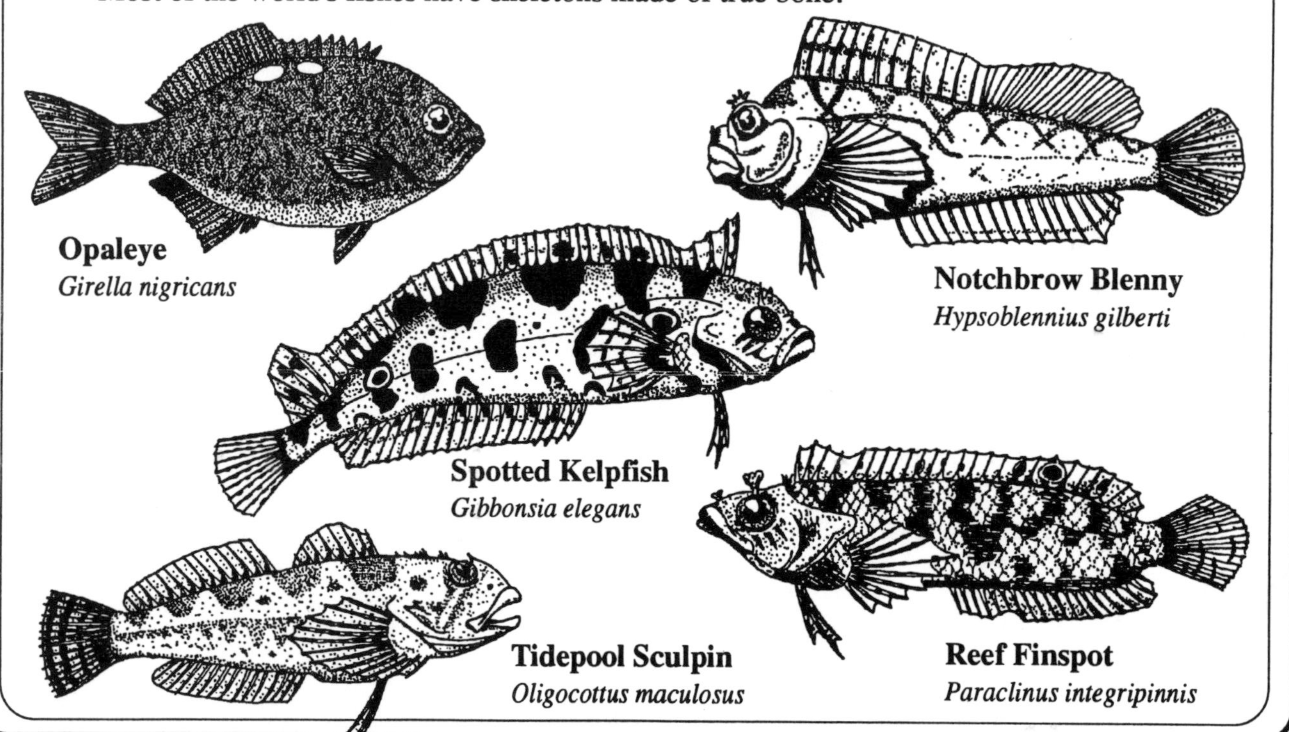

Opaleye
Girella nigricans

Notchbrow Blenny
Hypsoblennius gilberti

Spotted Kelpfish
Gibbonsia elegans

Tidepool Sculpin
Oligocottus maculosus

Reef Finspot
Paraclinus integripinnis

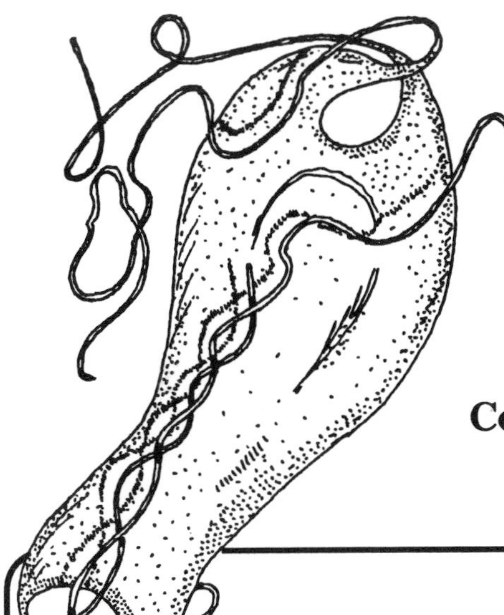

What's Inside a Mermaid's Purse?

Sometimes a strange, dark brown object (shown life size, left) washes up with strands of kelp. This is commonly called a "Mermaid's Purse," but it is really the egg case of a marine animal.

Connect the dots below to make a picture of the animal.

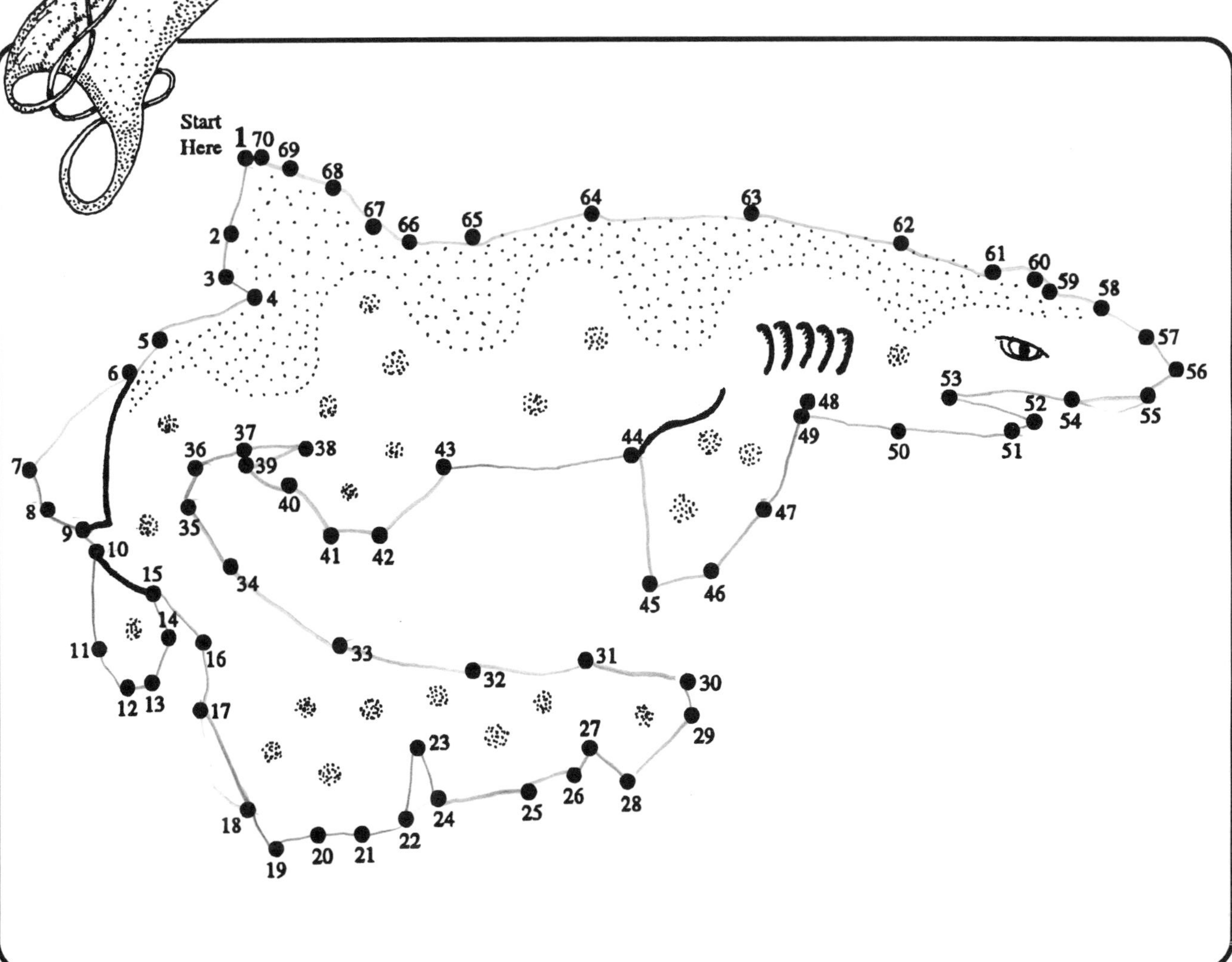

Assign numbers to the letters of the alphabet (1 = A, 2 = B, 3 = C, etc.). Then find the letters that match the numbers below to get the name of the animal you just drew.

S	W	E	L	L		S	H	A	R	K
19	23	5	12	12		19	8	1	18	11

Cryptic Coloration

Woolly Sculpin
Clinocottus analis

Lots of tidepool animals have colors or markings that help them hide from predators. This is called *cryptic* (KRIP-tick) coloration. Another name for it is *camouflage* (KAM-oh-flawdsh).

In the tidepool below are 12 more Woolly Sculpin fish just like the one above. Can you find them all?

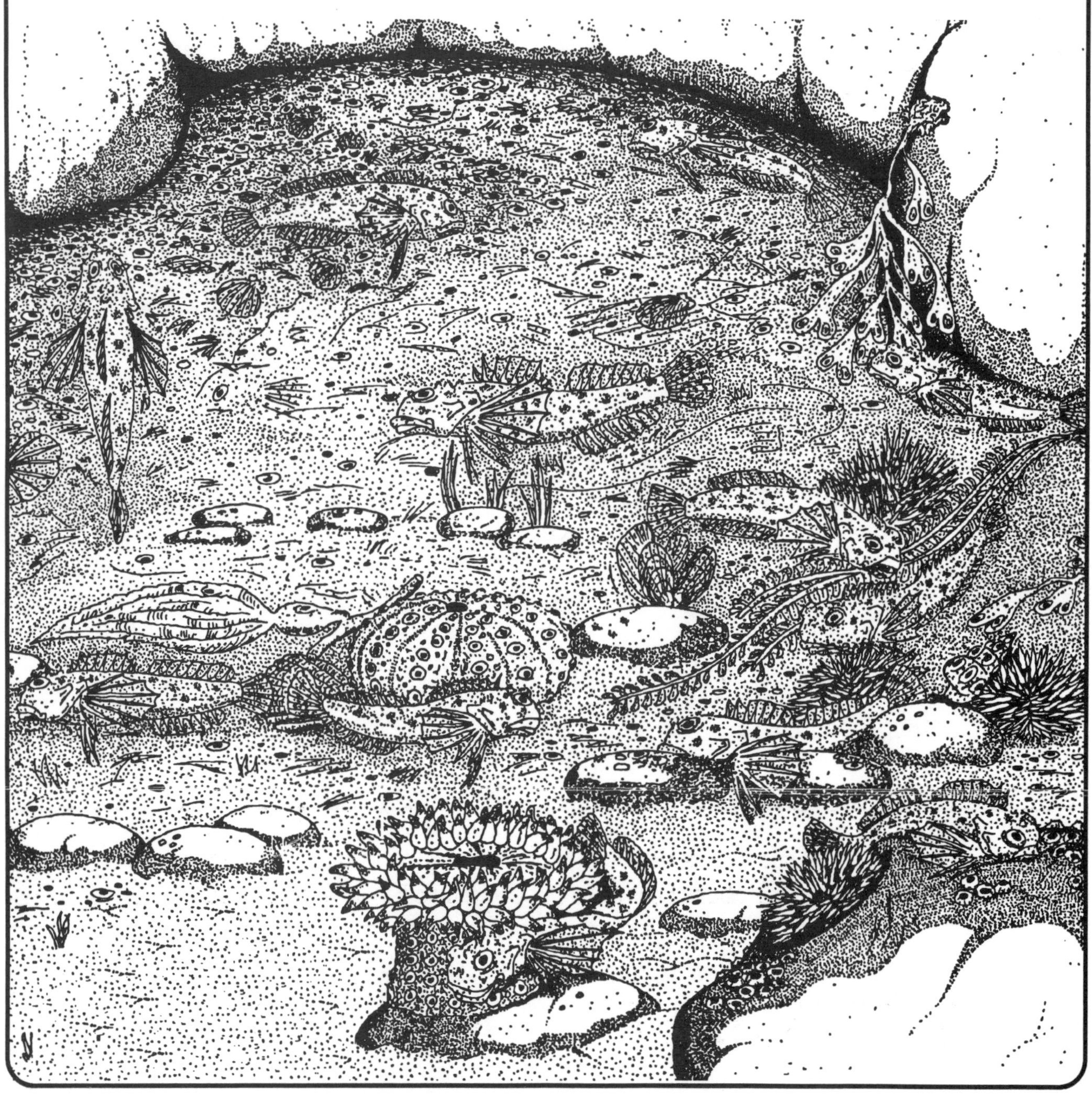

Debris in the Sea

Debris (deh-BREE) is another word for trash. For years, the ocean has been used as a dump for trash, sewage, hazardous waste, and pesticides. Plastics are the worst, because they don't break down into safe elements for many, many years. Here are some problems that plastics can cause:

Every year, millions of pounds of plastic fishing nets, buoys, lines and other gear is lost at sea. Whales, seals, and marine birds often become tangled in this gear and get injured or killed.

Sea turtles often feed on plastic bags, mistaking them for jellyfish. Many of these turtles eventually starve to death because the plastic clogs their digestive system.

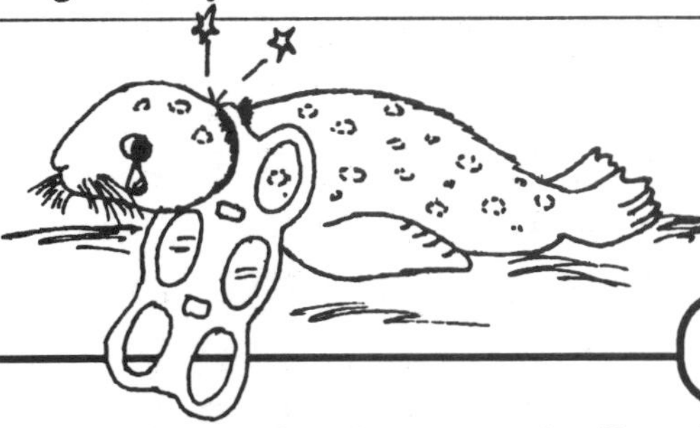

Young seals often play with plastic six-pack rings and get the bands caught around their necks. These bands can strangle them as they grow.

How YOU Can Help!

In 1988, an international treaty took effect that restricts plastic ocean dumping by the nations that ratified it -- including the United States. But there are still people who litter the beaches and illegally dump trash into rivers and oceans. **YOU** can help by doing the following:

- **Don't Litter**
- **Take Part in Beach Clean-up Days**
- **Be a Good Example to Other People**
- **Reduce-Reuse-Recycle**
- **Support Environmental Groups**

Hidden Treasures

Find the tidepool life listed at the bottom of the page in the letters below; they may be up, down, diagonal, forwards or backwards! Circle each name as you find it, and check it off on the list (one has been done to show you how).

```
G U M B O O T K E I G C E I B I A B A L O N E D
J N I E E H M J A C G N M D E O D B M Q K Q L Q
S P O N G E K F P M I R H S N M L R P P E J E J
C E L S W E L L S H A R K K O H C O E I O F S Q
F E I G B P K K H B P F A O M C I W J K W E S M
M T B A R C T I M R E H G G E K E N C L L A U S
H A B B F O M B D H A F C M N D N A B L L T M L
R C D P L I C C C B I O A D A J M L P E I H A O
A I B R I S T L E W O R M L E J H G U O M E I O
T N F P I P S E A H A R E F T D J A R B P R N P
S U G K O C O N O I K H D G A E P E P E E D R E
E T N O T K N A L P J L D O G C J E L L T U O D
L Q K S K H O C T O P U S D E Q O E E B A S F I
T I M C J A C K F R O S T B R Y O Z O A N T I T
T E J U N C Q Q B C K K S F G I B D L I G E L K
I E E L O P Q H B E K E B F G L E H I L M R A Q
R A L P T O G H L J A A G I A L M Q V M K W C O
B M L I I C B P F S R Q D A D O I B E L D O B P
F D Y N H P C E T N M G H N Q D O P O S I R P B
M E F N C R F A A B H C F E Q K J B H A N M D F
M L I J A M R C P B L E L A G B O L A H A A P E
G F S B I D L N K G S E A C U C U M B E R P O E
I B H F K E B S P I N Y L O B S T E R C Q J O N
N F L G S E I R E D S E A U R C H I N O Q D H M
```

Abalone
Aggregate Anemone
Barnacles
Bristle Worm
Brittle Star
Brown Algae
California Mussel
Chiton
Feather Duster Worm
Gumboot
Hermit Crab
Isopod
Jack Frost Bryozoan
Jellyfish
Kelp Crab
√ Octopus
Owl Limpet
Plankton
Purple Olive
√ Red Sea Urchin
Sculpin
Sea Cucumber
Sea Hare
Sea Star
Shrimp
Spiny Lobster
Sponge
Swell Shark
Tidepools
Tunicate

(Answer key is on inside of back cover)

Places to Visit

Tidepools occur all along the California coast, especially where there are large outcroppings of rock. Some of the very best tidepools are in protected areas such as parks and preserves. A few of these are shown on the map below.

In addition to the pools themselves, there are many places to learn more about marine life. Visitor centers, nature centers, campground exhibits, and museums can often provide you with information on the immediate area and about marine life in general. A few are listed on the next page.

Places to visit, continued

- **Natural History Museum,** Morro Bay State Park
- **Santa Barbara Museum of Natural History,** Santa Barbara
- **Sea Center,** Stearns Wharf, Santa Barbara
- **The Nature Conservancy**
  Guadalupe-Nipomo Dunes, Santa Maria and Guadalupe;
  Santa Cruz Island Project, Stearns Wharf, Santa Barbara
- **Visitor Center,** Channel Islands National Park, Ventura
- **Natural History Museum of Los Angeles County,** Exposition Park, Los Angeles
- **Cabrillo Beach Marine Museum,** San Pedro
- **Stephen Birch Aquarium-Museum,** Scripps Institute, La Jolla
- **Sea World,** Mission Bay, San Diego
- **Museum of Natural History,** San Diego

Books to Read

As you visit tidepools, you will find that you have more questions than answers! Where can you find the answers? In books! There are a lot of good books available on the southern California seashore. Here are some of them:

General References

Pacific Coast by Bayard H. McConnaughey and Evelyn McConnaughey. Audubon Society Nature Guide, Alfred A. Knopf Co., 1985.

The Marine Biology Coloring Book by Thomas M. Niesen. Barnes & Noble Books, 1982.

Between Pacific Tides by Edward F. Ricketts, Jack Calvin, Joel Hedgepeth and David Phillips. Fifth Edition, Stanford University Press, 1985.

Plants

Seashore Plants of California by E. Yale Dawson and Michael S. Foster. University of California Press, 1982.

Shells, Invertebrates

A Field Guide to Pacific Coast Shells by Percy A. Morris. Peterson Field Guide Series, Houghton-Mifflin Co. Second edition, 1966.

Seashore Life of Southern California by Sam Hinton. University of California Press, 1969.

Pacific Intertidal Life by Ron Russo and Pam Olhausen. Nature Study Guild, 1981.

Fish

Tidepool and Nearshore Fishes of California by John E. Fitch and Robert J. Lavenberg. University of California Press, 1975.

Pacific Coast Fish by Ron Russo and Ann Caudle. Nature Study Guild, 1990.

Tidepool Location Guide

Tidepools - Southern California by Linda E. Tway. Capra Press, 1991.